Politics of Technology in Latin America

This edited collection examines the political conditions affecting science and technology capability building in Latin America.

In a comparative discussion centring on Argentina, Brazil and Mexico, leading contributors look at the capacity Latin American governments have shown for building national systems for the support of innovation in industry. They look at how state capacities for the design and implementation of science and technology policies have evolved and examine how political factors, such as military rule and authoritarianism, have shaped such capabilities and the pattern of development. The influence of international factors on policy changes is also examined.

In looking ahead to the future economic and social development in Latin America, contributors find that selective but active state intervention in favour of technological change will be needed to ensure implementation of realistic technology policies.

Maria Inês Bastos is a Research Fellow at the UNU/INTECH. **Charles Cooper** is Director of the UNU/INTECH, The Netherlands.

UNU/INTECH Studies in New Technology and Development

Series editors: Charles Cooper and Swasti Mitter

The books in this series reflect the research initiatives at the United Nations University Institute for New Technologies (UNU/INTECH) based in Maastricht, The Netherlands. This Institute is primarily a research centre within the UN system, and evaluates the social, political and economic environment in which new technologies are adopted and adapted in the developing world. The books in the series explore the role that technology policies can play in bridging the economic gap between nations, as well as between groups within nations. The authors and contributors are leading scholars in the field of technology and development; their work focuses on:

- the social and economic implications of new technologies;
- processes of diffusion of such technologies to the developing world;
- the impact of such technologies on income, employment and environment;
- the political dynamics of technology transfer.

The series is a pioneering attempt at placing technology policies at the heart of national and international strategies for development. This is likely to prove crucial in the globalized market, for the competitiveness and sustainable growth of poorer nations.

1 Women Encounter Technology
Changing Patterns of Employment in the Third World
Edited by Swasti Mitter and Sheila Rowbotham

2 In Pursuit of Science and Technology in Sub-Saharan Africa
Edited by J.L. Enos

3 Politics of Technology in Latin America
Edited by Maria Inês Bastos and Charles Cooper

4 Exporting Africa
Technology, Trade and Industrialization in Sub-Saharan Africa
Edited by Samuel M. Wangwe

Politics of Technology in Latin America

Edited by Maria Inês Bastos and Charles Cooper

Published in association with the UNU Press

First published 1995
by Routledge
2 Park Square, Milton Park, Abingdon, Oxon, OX14 4RN

Simultaneously published in the USA and Canada
by Routledge

605 Third Avenue, New York, NY 10017

Routledge is an imprint of the Taylor & Francis Group, an informa business

Typeset in Times by LaserScript, Mitcham, Surrey

British Library Cataloguing in Publication Data
A catalogue record for this book is available from the British Library

Library of Congress Cataloguing in Publication Data
A catalogue record for this book has been requested
ISSN 1359–7922

ISBN 13: 978-0-415-12690-8 (hbk)

Contents

Part III Conclusion

Tables

Contributors

Maria Inês Bastos is a Brazilian political scientist with a PhD from the University of Sussex, England. She worked for five years as researcher at the Brazilian Council for the Development of Science and Technology (CNPq) and for six years as an adviser to the Brazilian Senate, particularly in the Committee on Science and Technology. In the latter position she collaborated in the development of the Brazilian Software Bill, the elaboration of the chapter on Science and Technology of the Brazilian Constitution, and the recent alteration of the Brazilian Informatics Law. She is presently a research fellow at the United Nations University Institute for New Technologies (UNU/INTECH), The Netherlands. Her research has concentrated on international and domestic political constraints on national policies for new technologies in developing countries. Her latest publication is *Winning the Battle to Lose the War. Brazilian Electronics Policy under US Threat of Sanctions* (London, Frank Cass, 1994).

Charles Cooper is the Director of the United Nations University Institute for New Technologies (UNU/INTECH), The Netherlands. For the past twenty-five years, Professor Cooper has conducted research, teaching and consulting in technology, economics and industrial development in developing countries. As Director of UNU/INTECH, he is responsible for planning and technical oversight for all research programmes. Professor Cooper served for ten years as Professor of Development Economics at the ISS, The Hague, and before that was for twelve years a joint fellow of the Institute of Development Studies (IDS) and the Science Policy Research Unit (SPRU) at the University of Sussex, England. His research has concentrated on technology transfer and the environmental and socio-economic impacts of technology and macroeconomic policy for development. His latest publication is *Technology and Innovation in the International Economy* (London, Edward Elgar, 1994).

Alejandro Nadal Egea is a Mexican social scientist with a Law degree from the Universidad Autonoma de Mexico and a PhD in Economics at the University of Paris, Nanterre. He has a professorship at the Centre for Economic Studies, El Colegio de Mexico where he co-ordinates the Science and Technology Programme. He has published widely in Spanish. His latest publications in English are 'Choice of technique revisited: a critical review of the theoretical underpinnings', *World Development* (vol. 18, no. 11, 1990), 'ICBM trajectories: some issues for the superpowers' neighbours', *Journal of Peace Research* (vol. 27, no. 4, 1990), and 'The development of Mexico's living marine resources', in S. Diaz-Briquets and S. Weintraub (eds) *Regional and Sectoral Development in Mexico as Alternatives to Migration* (Boulder, Col., Westview Press, 1991).

Fabio Stefano Erber is a Brazilian economist with a PhD degree in Economics from the University of Sussex. He was until very recently Director of the Brazilian Bank for Economic and Social Development (BNDES) and from the mid-to the late-1980s was the Deputy Secretary-General of the Brazilian Ministry of Science and Technology. In the latter position he organized the Brazilian National Development Plan for science, technology and industry and he designed and implemented specific policy measures for high-tech sectors and for higher education. He was full-time Professor at the School of Economics and the Institute of Industrial Economics (IEI) of the Federal University of Rio de Janeiro for ten years and was the Research Director of IEI for four years. He worked at the Research Department of the Brazilian Financial Agency for Studies and Projects (FINEP) of the Planning Ministry for eight years where he directed research projects on science and technology policy instruments, technological development of capital goods sectors and on state-owned enterprises. He has published mainly on industrial, scientific and technological development and state intervention in such areas.

Hugo Jorge Nochteff is an Argentine social scientist with an MA degree from the Facultad Latinoamericana de Ciencias Sociales (FLACSO). He is a senior researcher at FLACSO Argentina where he shares the directorship of the research programme on Electronics and Development in Argentina sponsored by IDRC in 1984–5 and, subsequently, by the Argentine National Council for Scientific and Technological Research (CONICET) and by the Volkswagen Foundation. He has been consultant to the Economic Commission for Latin America and the Caribbean (ECLAC), the United Nations Industrial Development Organization (UNIDO) and the International Labour Office (ILO) on matters of industrial policies, particularly for electronics and the impacts of new technologies on employment. He has published widely in Spanish on industrial policy and the development of electronics in Argentina.

José Nun is an Argentine political scientist with a Diplôme Supérieur from the Fondation Nationale des Sciences Politiques, University of Paris. He has been a professor at the Department of Political Science at the University of Toronto for tweny-four years, and a research fellow at the Center for Research on Latin America and the Caribbean, at the York University, Canada for seventeen years. He is the Director of the Latin American Centre for the Analysis of Democracy (CLADE) in Buenos Aires. Professor Nun participates on the board of editors of *Latin American Perspectives* (Los Angeles, CA) and *Labour, Capital, and Society* (Montreal, Canada). He has published widely in English and Spanish mainly on theoretical and historical analyses of authoritarian regimes, democracy and modernization in Latin America. His latest publication is *The Alfonsin Government and the Agrarian Corporations*, with Mario Lattuada (Buenos Aires, Manantial, 1991).

Preface

Maria Inês Bastos and Charles Cooper

Technology policy studies in Latin America lack systematic discussion on the political aspects of the process of decision-making and policy implementation. The present volume contains the result of a study designed to address this overlooked aspect. It deals with two major questions: (a) What capacity states in Latin America have exhibited to date for the promotion of science and technology activities? (b) How far the existence or lack of such capacity is to be explained by political factors? These questions oriented the case studies of Argentina, Brazil and Mexico.

The main conclusion is that the governments of these countries have shown strong capacity in support of innovation only through a few of their various agencies and in very specific industries. Beyond these sectoral experiences, the situation is not very positive. Most of what has been achieved in technology innovation is modest and was brought in by industrialization policies. This contrasts with the grand objectives of technology policy which were only partially attained. Two interconnected aspects of state–society relations help explain this result. One is the deficient or even conflictive relationship between state agencies and industrialists and engineers/researchers. The other is the political regime. In Latin America, authoritarianism has been not an asset but a liability for the implementation of realistic technology policies. Restoration of constitutional order, increased accountability and transparency create a more conducive environment for such policies. Focusing the state apparatus itself, administrative capacity and corporate culture were found to influence effectiveness in technology policy. Institutional weakness of technology policy agencies was tackled with centralization, but launched in a period of restrictions in public spending, it is not clear whether this was an efficient solution. Technology self-reliance, a central focus of the technology bureaucracy's corporate culture, became the main goal of policies such as those aiming at the development of local production capabilities in electronics and informatics. Their internal inconsistencies, the technology agencies' inability quickly to respond to changes in technology and society, and an eroding domestic support explain their eventual abandonment.

What role does science and technology policy play in future economic and social development in Latin America? Disenchantment with the results of state intervention has solid basis in regional history, but is also a by-product of the ideological offensive in favour of free markets. The unavoidable fact is that import substitution is dead and the discussion must therefore focus on technology policy in the context of liberalized open economies. A selective but active state intervention will still be needed. Competitiveness as a central objective of the current approach to liberalization can be attained under a variety of terms. Technology policy will not be a relevant issue where competitiveness is maintained mostly by a slow-growing structure of real wages. There is no example of a technologically based policy of competitiveness in developing countries in which the state has been neutral or inactive. The present Latin American experience suggests that when there is no serious intervention in favour of technological change, the economy tends to get stuck in traditional short-run patterns of comparative advantage. These may produce exports, but will they produce development?

Acknowledgements

We should acknowledge a number of debts. First of all to our contributors who took time from their various professional commitments to produce the papers for the workshop 'The Politics of Technology: Policy Institutions in Latin America'. The papers submitted for the workshop, which took place in Maastricht on 6–8 April 1993, were subsequently edited to provide the material for this book.

The discussants at the workshop made significant contributions to each individual paper under debate and the intellectual challenge of approaching the politics of technology in a region of more heterogeneity than is generally believed. We want to thank Lisa Bornstein, from BRIE-Berkeley University; Carlos Maria Correa, from Universidad de Buenos Aires; John Humphrey, from IDS-University of Sussex; Linsu Kim, from the College of Business Administration, Korea University; Joerg Meyer-Stammer, from the German Development Institute; and Samuel Wangwe, from UNU/INTECH, for their comments and suggestions on each paper.

We also take this opportunity to thank Bonnie Weinstein for the excellent copy editing and her commitment to work overnight and on weekends to ensure the authors met the publisher's deadlines. Sen McGlinn's creative insights helped in bringing the complex and elegant structures of the Spanish language into clean, simple and elegant English.

Finally, we are grateful to all our colleagues at UNU/INTECH for the superb intellectual environment where the anxiety of moving into the troubled waters of technology policy by less interventionist states has been faced with enthusiasm and courage.

John Humphrey, Samuel Wangwe, Carlos Maria Correa,
Joerg Meyer Stammer, Lisa Bornstein

1 A political approach to science and technology policy in Latin America

Maria Inês Bastos and Charles Cooper

This book deals with the political conditions for public-supported scientific and technological capability-building in Latin America. It focuses on explaining in which way science and technology (S&T) policies in individual countries evolve with time as well as how contrasting choices are made by governments of different countries, even though they face roughly similar economic and political situations.

The strategic role played by both technological change and capability in economic and social development is well established. They have become essential components of industrial strategies of many developed and developing countries. Recently this has been particularly true of science-based technologies such as electronics or information technologies (IT). In response to the microelectronics revolution, some countries have designed policies for fostering local technological capability in this area. In most of them, governmental efforts to take advantage of the opportunities opened up by the microelectronics revolution are related to a previously established, government-sponsored infrastructure of basic and applied research and also to the existence of state institutions for the promotion of S&T activities.

Within the developing world, experiences in government-sponsored promotion of science-based technologies such as microelectronics have varied in strategy and outcomes. In some newly industrializing countries, these policies are linked to an overall export-promoting industrialization based on imported technology. The result is an internationally competitive electronics industry. By contrast, in Latin America, state promotion of such science-based technology is launched in a context of import-substituting industrialization intended to build domestic design and production capabilities. The outcome of this alternative strategy is the creation of a domestically oriented electronics industry with a varied, though in general, low degree of international competitiveness.

The literature comparing 'outward-oriented' with 'inward-oriented' strategies of industrialization, or 'export-led' with 'import-substituting' growth, is abundant and in general praises the superiority of the outward/export orientation over the inward/import-substituting one. By implication, it is considered that export-oriented strategies contribute more significantly to technological development. The higher degree of international competitiveness of industrial products achieved through export-oriented strategies might be taken as testimony of the technological implications of this strategy since it assumes the implicit requirement of a higher technological content. This conclusion, however, needs some qualification. Much of the export performance of Latin American and many other developing countries illustrates that export competitiveness may sometimes be based less on improved technological capabilities and more on static comparative advantages accrued from natural resources and lower wages. It is important to stress that, while export orientation may stimulate technology capability-building, this is not a necessary and 'automatic' result stemming naturally from the strategy and, indeed, may not be the case at all. What is noteworthy is that export orientation is not an 'on-the-shelf' strategy to be picked up by rationally oriented governments. This is a point explored in this book: that technology policies, among other public policies, are a result of complex social and political processes which qualify the rationality of the choice of alternative options.

Efforts to understand the end results of these strategies have typically focused on the economic theories behind them and on their influence on the adoption of appropriate and coherent policies. Political analysis of these two strategies, however, draws attention to the sociopolitical conditions of policy choice and implementation. Governments are shown not to have the option of picking the technically best industrial strategy from a range of alternatives at their disposal, nor have they idealized administrative conditions to implement them. Policy choice depends not only on a government's capacity for obtaining and using relevant information, but particularly on its capacity for taking into account a complex set of social interests at the same time as keeping a degree of independence from social and political pressures. Policy implementation is directly dependent on administrative competences as well as on broad political support and consensus at various levels. Effective government intervention requires the building of state capacity for recognizing policy alternatives, mobilizing appropriate instruments, and developing the consensus necessary for policy implementation. Within the state apparatus a capacity for flexible adaptation is required as consensus may change and social support can dissipate owing to unforeseen international and domestic political developments.

The objective of the present volume is to contribute to a better understanding of science-based technology policy in Latin America by calling attention to the sociopolitical aspects of policy design and policy implementation. It also aims at forwarding a conscientious discussion of the technological and scientific capability-building by industrial policies in the region, either through direct or implied means. We believe that a more detailed analysis of the political conditions affecting sector-specific technology policies in Latin America will throw some light on positive and negative aspects of state promotion of technological development.

This chapter introduces the main topics discussed in the volume. Section 1 presents the objectives of the study that resulted in the various chapters of this book, sets the scope of the issues treated, and outlines each chapter. In Section 2, various conceptions of public support for S&T development are discussed, based on the experience of the governments of industrialized countries and of Latin America. Section 3 sketches an overview of influential Latin American literature on technological development and on S&T policy which argues in favour of a political approach. Finally, Section 4 highlights the explanatory potential of political factors that contribute or constrain S&T policy-making.

1 OBJECTIVES, SCOPE AND OUTLINE

A review of the many studies on S&T policy in Latin America indicates that systematic discussion on the political aspects of the process of decision-making and policy implementation is lacking. While many studies deal with the political economy of governmental intervention to stimulate innovation, there are few that relate how significantly the relationship between state and society in Latin America helps to explain chosen strategies. The present volume contains the results of a study designed to address this overlooked aspect of S&T policy in Latin America. The central issue dealt with by the various authors is state capacity for policy design, implementation and change. The analyses focus on the relationships between state and social interests and how they influence the formulation and implementation of policy as well as the organizational aspects within the state apparatus that either facilitate or constrain the state's capacity. Domestic institutions and international pressures are highlighted as facilitating or constraining forces explaining policy change.

Two sets of questions oriented the study:

1 How much capacity have Latin American governments shown for building national systems for the support of innovation in industry? How

have state capacities of design, implementation and change of S&T policies evolved in Latin America?
2 What role have political factors played in shaping this state capacity and to what extent can the evolution of such state capability be explained by political factors? In particular, what was the impact of changes in the political regime on the evolution of S&T policy in the region and how have international political factors influenced policy change?

These questions addressed two main objectives. First of all, focus was placed on policies designed to build general technological and basic research capabilities in a country, the core of 'national innovation systems'. The works conducted with this purpose in mind approached the governmental role in the formation of innovation systems by taking into account elements of macroeconomic policies, together with science policy and across-sectors technology policies. The analyses of policies for 'informatics' constituted the second part of the study. These policies were included in order to complement the picture of state capacity of design and implementation of S&T policy in the region. Elaboration on informatics policies illuminates the political conditions for state intervention, and its influences on the mastery and use of product design and manufacturing processes that are otherwise not only unknown but also close to international standard of 'best practice'.

The study intended to discuss the politics of technology policy in the Latin American region as a whole. Limitations of time and material resources, however, required the scope to be reduced to the three largest economies in the region: Argentina, Brazil and Mexico. These countries are included among those which have developed the most complex net of governmental institutions to deal with science and technology policies in Latin America. They have also made extensive efforts in formulating and implementing technology policies for informatics. While many of the issues discussed in this book may be relevant to the majority of countries in the region, the insights provided by the studies of the cases of Argentina, Brazil and Mexico cannot be generalized as to apply to the entire region.

The first part of the volume contains the works on the politics of building national systems of innovation in the three countries. In Chapter 2, Nun discusses the relationship between the social regime of accumulation and the political regime of government to explain both the absence of priority given to technology policies in Argentina and the successes and failures of them. The secondary role occupied by the autonomous development of science and technology in export orientation as it emerged in Argentina is illustrated in a comparison with Canada in relation to wheat production and exports. To explain the government's interest in fostering technical change in grain production, the author explores the decisive role played by the different class configurations and political systems which evolved in the two countries. The organization of a national scientific and technological complex that began in Argentina at the end of the 1950s issued from the government and not from the demands of the economy. Predictably, the government's efforts to establish links between the S&T complex and the sectors of production have had poor results. Relative failure is illustrated by the case of the Argentine INTI, the Institute of Industrial Technology, while relative success is exemplified by the cases of INTA, the Institute of Agrarian Technology, and CNEA, the Atomic Energy Commission. These two latter cases point to the developmental potential of the Argentine state in matters of science and technology policy.

The effects of a political regime on the capacity of the state for design and implementation of S&T policies is explored in all the works presented in this volume. In Latin America, and particularly in the three countries selected, explicit technology policies were primarily implemented under military rule or, as in the case of Mexico, under some form of authoritarianism. In addition, regime changes in Argentina and Brazil offer the possibility of further discussions of the relationship between political regimes and state capacities for S&T policy-making. The implications of authoritarianism for the implementation of policies to induce changes in the technological behaviour of firms in Brazil are discussed in Chapter 3, where Bastos explores the concept of 'embedded autonomy' of the state, or its relative insulation from society, as an important criterion for efficiency. While the import-substitution strategy of industrialization did not create economic incentives for innovative behaviour at firm level, the absence of institutionalized channels of communication with society, which characterized military rule in Brazil and has survived its overthrow, helps to explain the difficulties the government has faced in building an effective national system of innovation. Sectoral experiments combining state-induced research efforts and innovation performance of firms are examined as cases in which the Brazilian state showed strong capacity of policy design and implementation. These experiences, however, seem to be reduced to those in which the state is the main actor: cases of oil extraction, atomic energy and aeronautics. In these areas, there was nothing significantly obstructing the exchange of information between those who designed the policies, those who would have to implement them, and the firms that would have to innovate, since they were both component parts of the government or dependent on governmental demand.

The constitution of a national system of innovation in Mexico is analysed in Chapter 4 by Nadal. He views the lack of a democratic government and a juridical institutional framework as responsible for making policy design and implementation extremely vulnerable to manipulation by interest groups. Mexico's special form of authoritarianism resulted in a state whose strength is more illusive than real. The central goal of the state that had been, since the 1920s, to keep political stability in the country, has recently turned into maintaining PRI officials in office. This centrality of retaining political stability and its evolution to the stability in power of the major political party has had a negative impact on technology and industrial policies. S&T policies are shaped by a dominant linear conception of innovation in which the central role is played by basic research and the main actors are perceived to be academic researchers. The relationship between the scientific community and CONACYT (the central state agency of S&T policy in Mexico since 1971) is analysed to highlight, on one hand, the researchers' resistance to state priorities in the production of knowledge and, on the other, the use of science policy instruments by the government to control university researchers. This last issue became particularly relevant for political stability after the mass movements at the end of the 1960s. The relationship between the Mexican business world and the state is discussed in terms of attempts at manipulation by private interest groups and the vulnerability of state institutions to such manipulation. International incentives for policy change are illustrated by the discussion of two cases: (a) the ROTT (the Registry of Technology Transfer) in which international pressures were ineffective and (b) the change in the Mexican law for patents and trademarks, where international pressures were effective.

The second part of the volume concentrates on the informatics policies in Argentina and Brazil. In Chapter 5, Nochteff examines the electronics and informatics policy in Argentina between 1984 and 1988. This is considered the first, and up to now the only, public policy to have fostered science-based and specialized supplier industries in Argentina. It was also one of the few industrial policies to have been designed in Argentina since 1976. The policy has not only attempted to change the industrial structure but also the style of public policy in the country. The central argument developed in Chapter 5 is that it is precisely this attempt to reform public and private styles of policy and behaviour via a sectoral approach and with limited institutional means which has led to its limited results and eventual abandonment.

In Chapter 6, Erber discusses the evolution of the Brazilian informatics policy's objectives and instruments and the political alliances which supported and opposed the policy. He analyses the fate of the policy by restating the policy model and its technical and economic assumptions; the problems the policy had to meet in its attempt to manage the technological gap; and, finally, the values and interests at stake. Chapter 6 argues that the main attribute of the Brazilian informatics policy, i.e. development of a local innovation capability under control of national enterprises, made the policy intrinsically conflictive. Policy was designed based on various assumptions about the behaviour of entrepreneurs, the characteristics of the technology, and the consumers' reactions. Most of these proved to be wrong. While the strategists correctly assumed that multinational companies would not invest in R&D in Brazil, since they had the option to do so nearer their parent company in locations where they enjoyed strategic and systemic advantages, they were wrong when they assumed that national enterprises would invest the necessary amounts to absorb imported technology and develop their own technologies. Factors such as the speed and intensity of technological change and the threshold of resources necessary to catch up with the technological frontier and remain there were underestimated. The policy misplaced its emphasis on developing product design skills instead of production and marketing skills, and the policy-makers reacted very slowly to changes in the international environment. Social support of the policy shifted from co-operation and tolerance to hostility and civil resistance, which were nurtured by the media and international pressures. More than that, the author shows that the behaviour of entrepreneurs and consumers revealed a discrepancy between the dominant values of the Brazilian society and those implied by the policy.

Chapter 7 sums up the contributions of the case studies to answer the questions that oriented the study.

2 PUBLIC POLICIES FOR THE DEVELOPMENT OF SCIENCE AND TECHNOLOGY

S&T policies are forms of state intervention which not only promote the diffusion and development of knowledge as such but also incorporate practical application into production of goods and services. While treated as a single unit, they in fact have involved at least two different sets of policies – science policy and technology policy – each one, however, being subsumed under the other. This is true of Latin America and most of the countries that have dealt with this matter. The reason for approaching science and technology as a single policy area is the belief in the linearity of the innovation process, by which advances in basic research are thought to be necessarily followed by their incorporation in new products or production processes. Fostering the progress of science is believed to be the way to stimulate technological advance. The central institutional actors of such a unified policy area are the universities and the basic research organizations and to them the majority of the available resources are channelled.

The awareness of the many different roads taken by technological development, many with little connection to basic research, justifies the separation of technology policy from science policy. Innovation, or technological development, is a process of mastering and putting into practice techniques of design and production processes that are new to the firm or country (broader concept of innovation), or which are novel in absolute terms (narrow concept of innovation). Whatever the conception of innovation, tacit knowledge plays a central role in the process of 'learning by trying' and in producing incremental changes. This means that the central institutional actor of innovation is the firm, and technology policy intends 'to influence the decisions of firms to develop, commercialise, or adopt new technologies' (Mowery 1994).

Although the two areas have their specifics and the central institutional actors differ, they cannot be dealt with in complete isolation because of the many feedback loops that link the trajectories of the developing knowledge they produce. This is particularly true in relation to high technologies which are science based and need special skills for their use, not to mention their improvement and advance.

If the participation of each policy area in innovation cannot provide a clear criterion to separate them, can science and technology policies be distinguished from each other and from other policies by the instruments they use? This is another apparently simple question with a difficult answer. In the case of science policy, a wider consensus about policy instruments would be easier to reach than in the case of technology policies. Nobody would dispute that fellowship grants, scholarships and research grants, together with investment in the installation of research laboratories and libraries, and subsidies for importing research equipment, are among the instruments of science policy. Even mechanisms of communication between research institutes and industrial firms, despite some controversy this may provoke among those who prefer to separate the goals of basic research from its potential use, are nowadays accepted as science policy instruments. In relation to innovation policies, the big question is: what is to be included and what is to be left out – particularly when the concept adopted is the broad one (Nelson 1993: 5)? In this case, there might be a strong consensus for considering education and training policies together with direct support for industrial R&D or even the controversial protection to infant industries as instruments of technology policies. Great doubts would arise about including macroeconomic and stabilization policies which, however, do fundamentally affect firms' innovative behaviour. The distinction of what to consider as technology policies is hard to make and we would have to agree with the conclusion that 'One cannot draw a line neatly around those aspects of a nation's institutional structure that are concerned predominantly with innovation in a narrow sense excluding everything else, and still tell a coherent story about innovation in a broad sense' (Nelson 1993: 518).

Keeping this in mind, attention is focused on technology policies that express the clear intention of governments to affect the innovative behaviour of firms or, as it has been phrased in Latin America, of incorporating knowledge into production. It is well known that the distance between the intentions of governments and reality is often abysmal. It is, however, unavoidable to take the elusive characteristics of public policies into account if the analysis of policy choice is at stake. State capacity of technology policy design would be hardly grasped if attention were mainly centred on those policies that influence firms' decisions on innovation and adoption, but which 'have not been designed or implemented primarily to affect innovative performance' (Mowery 1994: 8).

S&T policies contribute to building 'national innovation systems' defined as networks of agents and sets of policies and institutions that affect the introduction of new technology to the economy; or as sets of public and private institutions and organizations that fund and perform R&D, translate the results of such activities into commercial innovations, and affect the diffusion of new technologies within an economy (Mowery 1994: 125; Dahlman 1990). National borders of innovation systems are being increasingly loosened by transnationalization, cross-country interfirm connections, and the imitation of organizational and management styles. Nevertheless, innovation systems still retain national peculiarities. This holds true not only for sector-specific national systems but also for a whole set of institutions that composes the S&T 'infrastructure', such as national educational systems, a nation's system of university research, public laboratories, and mechanisms for linking these research institutes and industry (Nelson 1993). On the other hand, one cannot expect to find much coherence and consistency in the set of institutions and policies that composes a national system of innovation. Nor can it be assumed that they result from a methodical effort to shape a consistent process of innovation. In this respect these institutions and policies do not constitute a 'system' in the strict sense of the word. In addition, the empirical experience of industrial and developing countries shows that such systems contribute to a generation of technological breakthroughs only on a sector-specific basis but are of central relevance when referring to innovation in a broader sense.[1]

There is another more politically explicit aspect of national innovation systems which is stressed in the works of the present volume. It has been a relatively recent concern of governments of industrial and developing countries to build such national systems of innovation. This concern has been justified by various reasons. 'National security' dominated the post-war period of the interest in high-tech areas. This has currently been overshadowed by 'national competitive advantages'. Some analysts seem not to be convinced of the national justification of such an effort and prefer to conceive technical advance as a product of the work of a community of actors irrespective of national borders (Nelson 1993: 15). Even though an extensive communication network links the research community on a worldwide basis, interfirm connections facilitate the diffusion of new technologies, and strategic partnerships are now extending to R&D activities, one cannot forget that technological advance remains the most powerful tool of competition (and economic rivalry) between firms and also between nations who sometimes question the policies of other nations on the grounds of 'fairness'. The works presented here do not substitute a political 'realist' view of technology advance for the idyllic conception of a worldwide communitary process of technical advance. They instead try to incorporate in the discussion the political contribution of international actors, including other countries' governments, in shaping national innovation systems.

2.1 Government support of S&T development

Throughout the world, government support and promotion have been an important part of the capitalist innovation system. In developing countries, S&T policies have precisely been designed to build up, almost from scratch in some sectors, or to model existing national innovation systems into a shape to fit economic and political goals. The roles played by government in this field have varied from country to country and throughout time.

Diverse forms of state action for the promotion of S&T have recently been analysed in terms of strategies or in relation to motives, objectives and instruments. Following Freeman's classification of strategies of firms, Niosi (1991) categorizes national systems of innovation as resulting from strategies which vary from offensive (the United States, Japan, Britain in the past), to opportunist (Sweden and Finland), to imitative (the present NICs), or defensive (European Union).

Roobeek (1990) emphasizes the motives, objectives and instruments of public policies. According to their motives, technology policies are classified as driven by concerns: of national security and free trade (the United States);

to secure raw materials and national sovereignty (Japan); to reinforce national competitive capability (West Germany); to improve the country's international competitiveness (France); to recover national competitive capability (United Kingdom); and to protect the 'universal welfare' idea (Sweden). Objectives of technology policies vary from keeping technological leadership (United States); gaining pre-eminence in the field of technology (Japan); catching up with the leader (West Germany); modernizing and restructuring the industry (France); becoming a standard-setter in fields of high technology (United Kingdom); and maintaining international competitiveness and restructuring industry to meet regional needs (Sweden).

In relation to instruments, technology policies have taken on different forms from massive R&D incentives, government contracts, incentives for private venture capital and deregulation in high-tech fields (United States); to the supply of information about foreign competitors, incentives for collaboration among large firms, and special support for core technologies (Japan). Added to those instruments some countries have extensive technology programmes of federal states, special loans, tax credits, and promotion of transfer of technology from research institutes to industrial firms (West Germany). While France has relied on R&D incentives to target areas, credit, price regulation, and support of medium and small firms, the United Kingdom has added to large-scale technology programmes, incentives to foreign investment, mobilization of venture capital, deregulation, and changes in the social security legislation. Sweden has based its policy on a favourable corporate climate, currency devaluation, tax credits, in addition to R&D funds (Roobeek 1990: 222–4).

The analogy between the strategy of firms and that of governments in their actions leading to national systems of innovation seems interesting especially because it draws attention to the central actor of technology development. However, it might be a little too artificial to think about complex, heterogeneous and often contradictory public policies, many of which are designed with other purposes, in terms of an overall government strategy. Although motives, objectives and instruments seem not to discriminate among different countries' experiences in technology policy, they are nevertheless criteria to distinguish policies directed towards influencing firms' innovative behaviour from policies directed to other ends. Even so, as mentioned before, it is difficult to draw a line between the policies that can be labelled 'technology' and others that have similar effects.

Mowery (1994) applies the well-known distinction between 'supply' policies yet distinguishes those promoting basic research infrastructure, which are very close to science policy (if basic research is mainly done in the universities) from those supporting industrial technology development.

In addition, he discusses 'adoption-oriented policies' which, in his view, mainly use the instruments of financial subsidies, provision of information and of extension services, regulation of technology transfer, definitions of technical standards, and government procurement. The conception of adoption-oriented policies is a clearer denomination for what has traditionally been referred to as technology policies, despite putting too much stress on a separation of activities that are hardly split in the real world. This limitation is fully acknowledged by the author who recognizes that the interaction of technology creation and technology adoption results in the inevitable overlap of adoption-oriented and supply-oriented policies.

A precise characterization of technology policies would need more than two criteria of classification. This was attempted by Kim and Dahlman (1992), who used a three-dimensional approach, integrating insights from literature on market mechanisms and technology flow with a dynamic perspective. According to market mechanisms, technology policies may be targeting the demand, the supply and/or the linkage between supply/demand. In relation to technology flow, policies may focus on acquisition, diffusion and/or the development of R&D activities. The dynamic perspective refers to the technological trajectory and distinguishes the emergency stage, the consolidation stage and the mature stage. A three-dimensional matrix represents an integrative approach and 'enables policy-makers to identify the areas that have been neglected and to assess the efficacy of policy instruments presently deployed' (Kim and Dahlman 1992: 441). The applicability of such an integrative approach is, however, closely dependent on the level of analysis as it relies on specific information of a technological nature.

For the purpose of the studies here presented, the various forms of state promotion of S&T were organized according to their intent, coverage, scope, organization and style. In relation to intentionality of government support, the promotion of S&T has varied. It jumps from the strategically conceived, clear and fully expressed purpose of improving science and technology to the implicitly received, as when state activity, serving other purposes, indirectly affects scientific and technological processes. As to the extent of its actions, government support has combined generality, when it is directed to strengthen research activities in any field, and selectivity, when it is significantly directed to the advancement of particular technology or a cluster of them, as in the case of 'new' technologies. According to their scope, government actions have been either narrow, when they are restricted to supporting research and development or to granting the supply of trained researchers, or comprehensive, when they also cover other innovative capabilities including quality control and managerial techniques.

There are two levels in which organizational aspects of government support of S&T have differed. As concerns the level of policy decision-making and implementation, S&T policies have been either centrally elaborated and implemented by a set of specialized organizations or designed and set in motion by organizations devoted to other activities, particularly in areas of defence and industrial policy. In relation to the dominant locus of government-supported S&T activity, policies have again varied from being highly centralized (governmental support concentrated on public research institutes or universities) to diffused (support mainly given to stimulate S&T activities at firm level). In addition, the style or mode of state intervention changed from strong, when governmental measures limit the choice of technology at firm level, to mild or weak, when government simply provides research infrastructures, including state-supported research institutes, or suitable environments for technology transfer.

The provision of education at all levels, support to R&D, regulation of technology transfer, adoption of any regime of protection of IPRs and its enforcement, trade policy, tax and financial incentives, competition policy, and government procurement, are among the major instruments governments have used to promote S&T development. The use of some of these instruments may be distinctive to some countries, but all of them have been used in varied composition in most of the countries. No clear pattern in the use of S&T policy instruments can be identified that could be explained by stage of development or the openness of the economy. The strongest evidence is related to the capacity for involvement of the private sector in research and development activities, which is consistently lower in developing countries.

The works that compose the present volume view S&T policy as a set of measures proposed and actually taken by governments acting intentionally in order to attain defined objectives, to generate and acquire knowledge and expertise, and to use them productively in the economy. The works explore variations in the forms of state intervention, the coverage and scope of the support, and the organizational structure across countries and throughout history.

2.2 Government support for S&T development in Latin America

The above-mentioned variations are also found among the countries of Latin America. S&T policy in the region has encompassed a set of measures for attaining the defined objectives of research development, technology transfer, and the use of science and technology for production and social welfare improvements. It has established new norms and regulations, allocated financial resources, and created organizations to implement policy and to co-ordinate related activities within the state apparatus. The various forms of explicit state intervention in S&T in Latin America can be described in relation to coverage, scope, organizational structure and style of state intervention.

S&T policy in Latin America has been directed to strengthening research in practically all fields of knowledge. Excellence of the research project, the need to create a critical mass of highly trained researchers, and the demand to reduce regional concentration of S&T have been dominant criteria for resource allocation. This picture is particularly illustrative of the period when government intervention in the field was science biased. Awareness among analysts and decision-makers of the specifics of the process of technology development and of the possible benefits of a more selective approach did not bring a significant change in the scope of S&T policy in the region. Nevertheless, and independent of this general approach, there have been significant cases of selective government promotion of science and technology research. Applied research and technological innovation have been supported by governments in selected fields such as agriculture (particularly for export crops), nuclear energy, small aircraft production, petroleum extraction and electronics.

The scope of S&T policy in the region can generally be described as narrow in the sense that it has almost exclusively supported research activities. Sector-specific technology policies have attempted to enlarge the scope of state intervention. However, despite being oriented towards more diversified forms of support, they have sometimes relied too much on the fact that technological accumulation could evolve automatically from production capacity. Without appropriate stimuli, manufacturing firms cannot create the capacity to generate and manage change. It seems that a more comprehensive approach to the process of innovation has been absent in S&T policies of both selective and general coverages.

There is little variation in the organizational structures of the major countries of the region. This structure was shaped, in a relatively short period of time, by the strong influence of international agencies. Since the mid-1960s, most of the countries in Latin America have had specialized state agencies, usually a 'Science and Technology Council', to take charge of the policy. These organizations did not have the power to do more than provide scholarships and allocate research grants to individual researchers or institutes. Organizational reform came about in the mid-1970s with the introduction of the concept of a national 'system' of S&T in which a central organization – the former council – articulates 'horizontally' and vertically through the various institutions and organizations acting in the field. These changes were meant to improve governmental effectiveness in the implementation of the S&T policy in the region. They were, however, fruitless, because the perceived inefficiencies were due to reasons other than organizational ones. In relation to the locus of S&T activity, public research institutes and/or universities have been the dominant beneficiaries of governmental funds for S&T activities in the region. Sector-specific technology policies have intentionally concentrated on strengthening R&D activities at firm level and establishing links between research institutes and production.

Government intervention in the region has been much stronger in economic policy than in S&T. Government-sponsored, import-substituting industrialization entailed setting up relatively coherent institutions for state intervention. These significantly included strong development banks, various ways of attracting and controlling foreign investment, controls for technology transfer, and the pervasive concept of 'local similar'. In contrast, government intervention for the promotion of S&T activities has mostly been confined to direct financial support of individuals or research institutes within the limits of a small S&T budget. Despite recognizing that disarticulation between economic and S&T policies has been the major source of mediocre results in the promotion of local S&T development and innovation, S&T policy-makers in the region have been unable to bridge the gap. A different picture can be drawn in relation to some sector-specific technology policies. Using the available instruments of economic policy, these technology policies have incorporated various degrees of intervention. They have supplied research in state-owned or state-supported institutes and transferred them at practically no costs to the private sector. As component parts of industrial policies, they have also sometimes imposed restrictions on company freedom of technology choice.

2.3 Evolution of S&T policy in Latin America

In a very broad sense and over and above specific characteristics of individual countries, the development of S&T policy in Latin America has been marked by changes in coverage, scope, organization and style of state intervention.

The origin of the policy can be located around the mid-1960s, but some organization-building was done earlier in the major countries of the region.[2] This has been characterized as one of 'imitative institutional development' (Bell 1985) because institutions of the industrialized world were transferred to the region without systematic analysis of the context into which they were transplanted. Moreover, there was no serious effort to design alternative institutional arrangements, and no attention was given to the complementary measures for generating technical change within the firms. It has also been considered as a clear example of the general development pattern of the region which was based on emulating the American experience (Adler 1987; Fajnzylber 1990). Until the mid-1960s, government initiatives in the area can be more appropriately characterized as science policy as they had mainly provided support to scientific research in public research institutes and universities. S&T policy in the region came out of the realization by decision-makers of the importance of planning the development of S&T according to a proper policy strategy that could organize government actions around defined objectives. This initiative was not completely domestically determined. The definition of national programmes for economic development was one of the conditions the countries in the region had to satisfy in order to get financial resources from the US government within the Alliance for Progress. In 1964, an OAS resolution urged Latin American governments to include S&T goals in their national development plans. The definition of science and technology policy was also based on a critique of the previous dominant belief that the development of science in itself would bring progress to the region (Mari 1982). Technology development gained prominence and became a policy target of its own. Local technological development, the adaptation of foreign technology, and control over the process of foreign technology transfer became components of the drive for science and technology self-determination. S&T plans were designed within the existing organizations that had been supporting scientific research. Therefore, the new concern with technology development did not significantly move the policy farther from what it had been doing before. S&T policy in the region has been, since its outset, general in coverage, narrow in scope, moderately centralized at the decision-making level, highly centralized in the support of public research institutes, and limited in government intervention. In the early 1970s, the acknowledgement of the need to focus attention not only on the supply of technology but also on its link to production led S&T decision-makers in the region to target the creation of 'effective demand' for science and technology. Conditions were ripe for a change towards a more selective and sector-specific approach. The acknowledgement of this demand justified government actions in some specific sectors. The organizational reform of the mid-1970s marked the end of this first phase.

In the early 1970s, the general coverage and narrow scope of S&T policy as well as its limited means of action were already under strong criticism. Two features were emphasized. First, articulation between S&T policy and economic policy and the integration of both into a global development strategy was lacking. Second, the general coverage and narrow scope of the policy were challenged. Many scientists, technologists and S&T policy decision-makers favoured concentrating efforts in specific sectors or projects as a way of establishing closer and more efficient links between firms, sectoral public agencies and finance institutions. This strategy change was seen as building up the national S&T system 'from below' (Mari 1982). Notwithstanding these suggested changes, the central action of the organizational reform of the mid-1970s was the creation of 'national S&T systems'. Not from below, but rather from above, was the combination of all institutions, organizations, agencies, societies and individuals directly involved in the establishment, structuring, development and operation of the scientific and technological activities in each country. These formally put together were considered the 'system'. S&T policy agencies (the councils) placed at the head of this structure carried out the functions of planning, articulating and co-ordinating the whole. The second phase of S&T policy in the region was marked, therefore, by a change in government strategy. Instead of a shift in orientation of the S&T policy itself, it not only brought a clear separation between 'real' technology policy – which definitely became sectoral – and the merely formal S&T policy, but it also passed on some of the biases of the general S&T policy to sectoral technology policies. Decision-making in S&T policy was made more centralized, at least in formal terms. State intervention in S&T policy with general coverage did not change its intensity, while some sectoral technology policies in the region reached higher levels of interventionism. The achievements of sector-specific technology policy were accompanied by the significant loss of credibility of S&T policy by its direct clientele – academic researchers – and the 'real' actors of technology policy-industrial entrepreneurs and other government segments. At the end of the 1980s, the significance of national S&T policy for innovation systems in the region was rather small and the S&T policy central agency was an empty shell.

The economic crisis and stabilization programmes in the 1980s hit the S&T policy in the region hard. Scarce financial resources during the decade decisively affected many research programmes and institutes. The organizational head of a national S&T system was among the areas affected by the reduction in the size of public administration. This resulted in the dispersion of government cadre in S&T policy. The quality of higher education became affected not only by scarce resources but also by a deterioration in basic and intermediate education. In most Latin American countries the deterioration of the S&T infrastructure has been identified as the most important problem to be tackled in the 1990s. In line with the programmes of liberal governments currently dominant in the region, actions to restore the S&T infrastructure have been proposed on a basis of freedom of choice. Research institutions have the power to set out their own priorities and technology development is left to market forces. The effects of the overall liberal orientation of the governments in the region on sectoral technology policies were immediate and clear. Those policies containing strong elements of interventionism were subject to major adaptation.

A more accurate characterization of the evolution of Latin American S&T policy is one of the achievements of the works presented in this volume. The brief description made here is sufficient to show the reader that important changes have occurred in the S&T policy in the region and that a better understanding of such changes is to be sought.

3 UNDERSTANDING TECHNOLOGICAL DEVELOPMENT AND S&T POLICY IN LATIN AMERICA

There is voluminous social sciences literature written on S&T activities in Latin America. The literature written in the mid-1960s and later reviewed by Oteiza (1991) mostly includes economic and sociological studies on the determinants of technology change, the nature of domestic technological effort, and the social consequences of technological innovation. The literature produced in the mid-1980s and reviewed by Schwartzman (1989) and James (1993) explores the role of state intervention in the acquisition, generation and incorporation of technology, the roles played by other relevant actors in these processes, and the convenience or extent of a developing country's efforts to participate in the new technology revolution. Analyses of Latin American S&T policy itself constitute a much smaller set of literature. In the case of Brazil, there is a thorough review of studies on S&T policy, but it is now more than fifteen years old (Erber 1979). Much of the literature on Brazilian technology policy concentrated, during the 1980s, on the discussion of the pros and cons of the Brazilian informatics policy. However, the review of this copious literature is yet to be done.

Several studies have treated the implications of the political economy of underdevelopment in relation to the technological lag and reduced innovative capability in the region. Works by Herrera (1968, 1971) lead a copious amount of literature analysing the technological lag from a historical and structural approach to development, the so-called 'Cepalian model' [CEPAL is the Spanish acronym for ECLA], which is used in a more radical way by Varsavsky (1969, 1972). Identifying the structural roots of the scientific and technological lag, they have shared the idea that a change in society would be necessary in order to eliminate all forms of underdevelopment, including scientific and technological. The historic and structural approach has also influenced empirical studies such as those on the availability of engineers and the 'brain drain' by Sito and Stuhlman (1968), Slemenson *et al.* (1970) and Oteiza (1971a, 1971b). These studies help to illustrate constraints put on efficient investment during the generation of scientific and technological capability in the region.

Another set of social studies on S&T activities in Latin America evolved from the Cepalian model into the dependency approach. Works by Sabato and Botana (1968), Sunkel (1970), Sabato (1971) and Monza (1972) among others are representative of this stream of literature. Technological dependence was expressed by a lack of articulation among key actors involved in the promotion of research, production of knowledge, and use of technology in Latin America. It was concluded that there is no real S&T system in the region, the expression being only a formal façade of developed institutions masking the real situation of peripheral countries. Sabato's 'triangle' model was used to show not only the weakness of S&T infrastructure, governmental inability of integrating S&T policies within development strategies, and low innovative capability at firm level, but also the lack of links between these three actors of S&T activity: research institutes, government and firms. The consequences of import-substituting industrialization and of the international division of labour for technology development in Latin America were initially analysed by Sunkel (1970). He showed that the central role given to subsidiaries of foreign companies whose R&D activities were not transferred to the host countries reinforced Latin America's position as mere user of the technology generated in the developed world. Extending this argument, Monza (1972) among others pointed to the specifics of the process of technological change in Latin America which he found could be better explained by income distribution, import substitution, and the pattern of foreign investment than by the relative costs of capital and labour.

The process of technology import and absorption in Latin America was clarified by the works of Vaitzos (1971), Katz (1971, 1976) and Wionczek and Leal (1972). They influenced the design of institutions, policies and instruments for the control of technology transfer. Contrasting with the conclusions of the dependency-oriented studies, Katz (1976) has shown that capacity for adaptive industrial innovation is not lacking in the region. More recently (Katz 1982, 1987) he has proved that technology generation by firms is a gradual learning process. The intensity and the direction of this effort depends on the competitiveness of the environment and the broad parameters of the system in which the firm and industry are immersed. And if the external and institutional constraints are not too limiting, the firms which had engaged in the learning process can approach the international levels of productivity. Government policies have a significant role in inducing technology choice by firms. This occurs mainly when intervention in the competitive environment affects the technological search effort within firms (Katz 1982, 1987).

Social consequences of technological change in Latin America have long since been analysed by Versiani (1971), Araujo and Pereira (1976), and Acero (1982, 1983) among others, who pioneered the literature on the impact of technological innovation (particularly automation of industrial production) upon employment, wages and working conditions at shop-floor level. Those studies have shown that the immediate results of technological change are not favourable to the Latin American working class. This is illustrated by the fact that increased automation in the textile industry caused redundancy, workers employed before the introduction of the new production process could no longer practise their skills, the number of skilled jobs did not increase, and control over workers' activities was strengthened.

Finally, examples of studies on S&T policy in Latin America until the late 1970s can be found in Erber's review of the Brazilian literature (1979: 163–70). According to him, historical analyses show that up to the late 1960s, S&T policy had not been a subject for systematic intervention by the Brazilian state. In the late 1970s, Erber identified a lack of detailed assessments of S&T policy outcomes. He linked this to the fact that the period of the policy implementation had been too short for any outcome to have matured. Even so, the literature pointed out that structural limitations of such policy stem from its objective of reducing technological dependence, which contradicted the overall strategy of development in Brazil: while explicit S&T policy pushed for an increase in technological autonomy of Brazilian firms, other policies induced domestic entrepreneurs increasingly to make use of foreign technology.

Analysing the studies on the launched Brazilian S&T policy, Erber identifies three points of consensus. First, the pattern of development followed in Brazil as well as the implicit S&T policy (including S&T policy implications of the second phase of import-substituting industrialization) are seen as structural limits for an explicit S&T policy, constraining the scope for effective intervention to a narrow list of products. Second, an assessment of the explicit S&T policy shows a continual low level of efficiency and inconsistency with the implicit policy. To achieve correlation, both policies (implicit and explicit) have to be designed at industry level so that they can take into account specific features of innovation in each industry. Finally, it is agreed that time dimension is of great relevance for S&T policies as their outcomes require long maturation. This implies that S&T policy-makers have to be able to anticipate problems and opportunities.

At the time of Erber's review, Gutierrez and Blanco (1980) completed a study of the practice of S&T policy in Venezuela. This study, especially in respect of the analysis of the two plans for science and technology, illustrates the fact that the Brazilian case had much in common with other countries in the region. Gutierrez and Blanco acknowledged the lack of political analysis of S&T policy in the region and contributed to filling in this gap by identifying the relevant groups and discussing their roles in S&T policy-making in Venezuela. They characterize the following players: the scientists; the fractional group of engineers acting towards the generation of technology and the rendering technological services; industrial producers, especially those in leadership positions in manufacturing and in corporative associations; the planning community of agencies in charge of S&T policy; the state bureaucracy; and the politicians, a group that includes the professional politicians and those who occupy top positions in government. They conclude that there is no science policy in the country nor is there a technology policy. The latter is only an ideological verbalization and in both cases the explicit policies represent only the illusion of planning. The practice of S&T policy in Venezuela is described in terms of:

1 confusing policies for the development of science with policies for technological development;
2 retaining an unintentional, despite formal rejection of, science bias;
3 being predominantly supply oriented;
4 approaching the supply of S&T as a matter of supporting R&D;
5 setting goals which are a mere reflection of perceived requirements of national plans;
6 superficially defining concrete conditions for implementation.

Unfortunately the authors have not systematically used the information they had on the relevant actors in Venezuelan S&T policy to provide an explanation of the evolution of such policy.

Understanding high-technology policies in Latin America was the objective of Adler (1987), Cline (1987) and, with an international comparative perspective, of Nau (1986) and Ramamurti (1987). All those works were reviewed by Schwartzman and James. Adler studied the Argentine and Brazilian development of computer and nuclear energy industries shedding some light on the possible explanations of technology development in cases where structural indicators would have shown only a small potential for it. His analysis emphasizes the role played by the pragmatic antidependency theory as a model for shaping reality and supplying the know-how, the 'know-what', and also the 'know-where-to' for S&T policy-makers. This theory provided the ideological basis for the 'antidependency guerrillas' – an alliance of combative state officials and academic researchers – who fought against the traditional 'fracasomania' in Argentina and the established open-door policies regarding multinational corporations in Brazil, and who initiated the two technology policies in those countries.

The Brazilian computer policy taken by Adler as a case of a success story is not approached in such enthusiastic terms by Cline (1987), who stresses the high costs involved in the building of a domestic computer capacity in the country and by Nau, (1986) who sees it as a big failure. For Nau, and a little later Wade (1990), among many others, the important question is what kind of policy is more successful and not whether governments play a role in the development of new technologies. Import-substituting policies, as the Brazilian informatics policy, are viewed as less successful than those oriented to the domestic/export market, as in South Korea and Taiwan. Schmitz and Cassiolato (1992) seem to bring to a close the long and heated controversy about the Brazilian informatics policy, showing its strengths and weaknesses, stressing the learning capability to take informed decisions that the policy helped to build in both industry and government in Brazil, and discussing the economic and political conditions under which such learning occurs. Ramamurti (1987) studied successful and less successful cases of state-owned enterprises in high-technology industries in India and Brazil and found that appropriate macroeconomic circumstances together with managerial autonomy, entrepreneurial attitude, and a strategic mix of imported technologies and research and development efforts, explain success in making products that require sophisticated technology. In any case, two fundamental questions seem to remain unanswered. First, why did some countries choose one strategy and others a different one? Second, what are the internal political conditions that stimulate a country's ability to seize the changing technological and economic opportunities in the world?

4 AN APPROACH TO THE POLITICS OF TECHNOLOGY IN LATIN AMERICA

Most of the literature we have discussed dates from the period of import-substitution, which covered both the first and second phases of science and technology policy in Latin America. There is much less written which expressly addresses the questions of technology policy under the aegis of trade liberalization. In part this book is intended as a contribution to the post import-substitution discussion.

The broad outlines of the change are clear to see. Import-substitution policies had their own economically and politically powerful constituencies. Elements of the industrial owning classes, owners of financial capital, middle-class management and some nationalistic groups could make common cause with more or less progressive nationalist political groupings imbued with the ideas of dependency thinking in support of a nationally oriented industrial strategy. The need for *national* science and technology capabilities and an adequate national institutional framework for science and technology, as an integral part of this domestically oriented development strategy, was taken more or less as given. It was easy for important elements of the scientific and engineering communities in Latin American countries in the 1970s and early 1980s to ally themselves with the forces supporting the import-substitution project. The domestically oriented science and technology policy which this project dictated, and the elements of technological protectionism which it entailed, were probably congenial to important parts of the Latin American science and technology communities. And the broad alliances between science and technology on the one hand and the economic and social forces underlying import substitution as an economic strategy on the other, were the basis for the technology policies of the time. These were the alliances behind the second phase of science and technology policy in many countries, the forces which supported the CONACYTS, the technology registers, the policies of learning by doing in the local engineering design sectors, and the widespread use of state interventions to induce the growth and application of local technological capabilities through what amounted to a kind of technological infant industry approach.

The broader international and national economic developments that led to fiscal crisis, and finally to the debt crisis and the unsustainability of the import-substitution project, also brought these key political alliances, which underlay the whole structure of policy, to an end. There is yet to appear a fully satisfactory account of how the alliances split apart and reformed around a totally new and quite different economic programme. For the moment one can only speculate about some of the factors that might have been involved. Plainly, the older alliance reached a point of deep contradiction. Some parts of the picture are apparent. For example, attempts to sustain import-substitution policies by moving to deeper levels of industrialization in the producer goods sectors strained the community of interests in the industrial sector to breaking point. For those who might benefit from the new levels of protectionism as producers of capital goods, such new developments were no doubt welcome. For their customers, the firms which would make use of the domestically produced capital equipment, the picture might have been less attractive, especially if – as must often have been the case – the new equipment was less productive, and less profitable at the existing structure of prices for final output than was the old. And such equipment users might also have looked without enthusiasm at technology policies that restricted reliance on foreign technology in the capital goods sector in order to encourage the growth of local capabilities. In short, the economic and political oppositions aroused by the deeper levels of import substitution, which by the 1980s were the logical next step, were almost certainly more severe than at any earlier stage. And these oppositions had powerful implications for technology policy as well, and no doubt weakened the once-powerful alliances between the science and technology communities and the now-divided industrial community. It should not be surprising then that the consensus around the desirable objectives of technology policy broke down.

What has come to replace the older alliances? There is at present no very clear answer. In a priori terms, one might imagine one of two possibilities. The first is that an industrialization pattern will emerge based on very static, short-run comparative advantage considerations, determined strictly by today's relative prices. This will move accumulation in the industrial sectors towards technologically static sectors, natural resource dominated in all probability, and where a relatively low and slow-growing structure of real wages will be sufficient to maintain competitiveness. Technology policy will not have an important role in such a pattern and the economic role of scientists and technologists in this kind of industrial system would presumably be rather limited. In this situation the political alliances around technology policies of earlier years would dissipate in a definitive way.

The second alternative is that there will be a focus on industrial sectors that are comparatively dynamic in technological terms and in which competitiveness will require access to foreign technology in order to maintain the position of firms in international markets. In these sectors, the liberalized world economy has made access to foreign technology a more important consideration than it was in the import-substituting system, since it is now a matter of competitive survival for firms and industries. This is the kind of pattern that came progressively into existence in the Southeast Asian economies. It is associated with an activist – though highly selective – role for the state and it is based on an active and strong alliance of national capitalists seeking to use the local science and technological system as a factor in the international competitive struggle in which they are engaged. This is a style of open-market development policy in which there is a strong political demand for local science and technology, stronger in all likelihood than under the auspices of the import-substitution strategies of earlier years (though conceivably able to build on the accumulated capabilities and the institutions left behind by the import substitution experience). It is not at all clear which of these alternatives will predominate in the new economic and political configurations of Latin America, or indeed whether some other may not emerge. Some of the observations and analysis in this book contain pointers and we will return to these later at an appropriate point.

NOTES

1 See the recent book edited by Nelson (1993) where the cases of industrial economies and developing countries are fully analysed.
2 Creation in the 1950s, with UNESCO advice, of S&T councils in the largest countries in Latin America. The Mexican Instituto Nacional de la Investigación Científica (National Institute for Scientific Research) was created in 1950; the Brazilian Conselho Nacional de Pesquisas (National Research Council) in 1951; and the Argentine Consejo Nacional de Investigación Científica y Tecnológica (National Council for Scientific and Technological Research) in 1958.

REFERENCES

Acero, L. (1982) *Impact of Technical Change and Traditional Skills: the Textile Sector in Brazil*, Final Research Report, IDRC/IUPERJ, March.
Acero, L. (1983) *Technical Change, Skills and Labour Process in a Newly Industrializing Country: A Study of Brazilian Firms and Workers Perception in the Textile Sector*, D. Phil. dissertation, School of Arts and Social Studies, University of Sussex.
Adler, E. (1987) *The Power of Ideology. The Quest for Technological Autonomy in Argentina and Brazil*, Berkeley, University of California Press.
Araujo, J. T. and Pereira, C. M (1976) 'Teares Sem Lançadeira na indústria têxtil', in IPEA/INPES *Difusão de inovaçoes na indústria brasileira: tres estudos de caso*, Rio de Janeiro.
Bell, M. (1985) *The Great Experiment: Harnessing Science and Technology to Third World Development. A Review of Policy and Policy Analysis since the 1950s*, Brighton, University of Sussex, SPRU.
Cline, W. R. (1987) *Information and Development: Trade and Industrial Policy in Argentina, Brazil, and Mexico*, Washington, Economics International.
Dahlman, C. J. (1990) *Electronics Development Strategy. The role of government*, World Bank, Industry and Energy Department Working Paper, Industry Series Paper no. 37.
Erber, F. S. (1979) 'Política científica e tecnológica no Brasil: uma revisão da literatura', in João Sayad (ed.) *Resenhas de Economia Brasileira*, São Paulo, Editora Saraiva.
Fajnzylber, F. (1990) 'The United States and Japan as models of industrialization', in Gary Gereffi and Donald L. Wyman (eds) *Manufacturing Miracles. Paths of Industrialization in Latin America and Asia*, Princeton, NJ, Princeton University Press.

Gutierrez, I. A. and Blanco, M. A. (1980) *La planificación ilusoria. Ensayo sobre la experiencia venezoelana en política científica y tecnológica*, Caracas, Cendes/ Editorial Ateneo.

Herrera, A. (1968) 'Notas sobre la ciencia y la tecnología en el desarrollo de las sociedades Latinoamericanas', *Revista de Estudios Internacionales*, Universidad de Chile, año 2, no. 1, Santiago.

Herrera, A. (1971) *Ciencia y política en America Latina*, Mexico, Siglo XXI.

James, D. D. (1993) 'Technology policy and technological change: a Latin American Perspective', *Latin American Research Review*, vol. 28, no. 1, pp. 89–100.

Katz, J. M. (1971) *Importación de tecnología, aprendizaje local e industrialización dependiente*, Buenos Aires, Instituto Torcuato Di Tella.

Katz, J. M. (1976) *Importación de tecnología, aprendizaje, industrialización dependiente*, Mexico, Fondo de Cultura Economica.

Katz, J. M. (1982) 'Technology and economic development: an overview of research findings', in Moshe Syrquim and Simon Teitel (eds) *Trade, Stability, Technology, and Equity in Latin America*, New York and London, Academic Press.

Katz, J. M. (1987) 'Domestic technology generation in LDCs: a review of research findings', in Jorge M. Katz (ed.) *Technology Generation in Latin American Manufacturing Industries*, London, Macmillan.

Kim, L. and Dahlman, C. J. (1992) 'Technology policy for industrialization: an integrative framework and Korea's experience', *Research Policy*, vol. 21, pp. 437–52.

Mari, M. (1982) *Evolución de las Concepciones de Política y Planificación Científica Y Tecnológica*, Washington, Programa Regional de Desarrollo Cientifico y Tecnologico, Departamento de Asuntos Cientificos y Tecnologicos, Secretaria General de la Organización de los Estados Americanos.

Monza, A. (1972) 'La teoria del cambio tecnológico y las economias dependientes', *Desarrollo Economico*, vol. 12, no. 46.

Mowery, D. C. (1994) *Science and Technology Policy in Interdependent Economies*, Boston/Dordrecht/London, Kluwer Academic Publishers.

Nau, H. (1986) 'National policies for high technology development and trade: an international and comparative assessment', in Francis W. Rushing and Carole Ganz Brown (eds) *National Policies for Developing High Industries, International Comparisons*, Boulder, Col. and London, Westview Press.

Nelson, R. R. (ed.) (1993) *National Innovation Systems. A Comparative Analysis*, New York/Oxford, Oxford University Press.

Niosi, J. (1991) 'Canada's national system of innovation', *Science and Public Policy*, vol. 18, no. 2, pp. 83–92.

Oteiza, E. (1971a) 'Emigración de profesionales, tecnicos y obreros calificados de Argentina a los Estados Unidos: analisis de las fluctuaciones de la emigracion bruta, julio 1950 a junio 1970', *Revista de Desarrollo Economico*, Buenos Aires, vol. 10, nos 39–40.

Oteiza, E. (1971b) 'Un replanteo teórico de las migraciones de personal altamente calificado', in Walter Adams (ed.) *El Drenaje de talentos*, Buenos Aires, Paidos.

Oteiza, E. (1991) *Los Estudios Sociales de la Tecnología en la Region Latinoamericana. Diagnóstico y Perspectivas*, Buenos Aires, Centro de Estudios Avanzados, Universidade de Buenos Aires, Serie Documentos 2/91.

Ramamurti, R. (1987) *State-owned Enterprises in High Technology Industries: Studies in India and Brazil*, New York, Praeger.
Roobeek, A. J. M. (1990) *Beyond the Technology Race. An Analysis of Technology Policy in Seven Industrial Countries*, Amsterdam, Elsevier.
Sabato, J. (1971) *Ciencia, tecnología, desarrollo y dependencia*, Universidad Nacional de Tucuman.
Sabato, J. and Botana, N. (1968) 'La ciencia y la tecnología en el desarrollo de America Latina', *Revista de Integracion*, no. 3.
Schmitz, H. and Cassiolato, J. (eds) (1992) *High-tech for Industrial Development. Lessons from the Brazilian Experience in Electronics and Automation*, London and New York, Routledge.
Schwartzman, S. (1989) 'The power of technology', *Latin American Research Review*, vol. 24, no. 1, pp. 209–21.
Sito, N. and Stuhlman, L. (1968) *La emigración de científicos de la Argentina*, San Carlos de Bariloche, Fundacion Bariloche.
Slemenson, M. (1970) *Emigración de científicos argentinos: organización de un exodo a America Latina. História y consecuencias de una crisis universitária*, Buenos Aires, Instituto Torcuato di Tella.
Sunkel, O. (1970) 'La universidad latinoamericana ante el avance científico tecnico', *Revista del Instituto de Estudios Internacionales de la Universidad de Chile*, año 4, no. 13.
Vaitzos, C. (1971) 'Opciones estratégicas en la comercialización de tecnologia: el punto de vista de los países en desarrollo', *Revista Comercio Exterior*, Mexico.
Varsavsky, O. (1969) *Ciencia, política y cientificismo*, Buenos Aires, Centro Editor de America Latina.
Varsavsky, O. (1972) *Hacia una política nacional*, Buenos Aires, Ediciones Periferia.
Versiani, F. R. (1971) *Technical Change, Equipment Replacement and Labor Absorption: The Case of the Brazilian Textile Industry*, PhD Thesis, Vanderbilt University.
Wade, R. (1990) *Governing the Market. Economic Theory and the Role of Government in East Asian Industrialization*, Princeton, NJ, Princeton University Press.
Wionczek, M. and Leal, L. (1972) 'Hacia la regionalización de la transferencia de tecnologia en Mexico', *Comercio Exterior*, Mexico.

Part I

The politics of building national systems of innovation

2 Argentina: science, technology and public policies

José Nun[1]

Two sets of questions of undisputed importance oriented the project of which this study is a part: on the one hand, what are the capabilities exhibited so far by various Latin American states in promoting science and technology (S&T) policies; and, on the other hand, how are these capabilities, or absence thereof, explained? Any attempt at answering such questions, particularly those of the second type, requires that they first be placed in a certain theoretical context – an operation that is always full of consequences since, among other things, it determines the specific level of analysis at which the answers are sought.

Thus, as Lindblom (1977: ix) already warned some years ago, when 'political science turns to institutions like legislatures, civil service, parties, and interest groups, it has been left with secondary questions'. Why? Because 'the operation of parliaments and legislative bodies, bureaucracies, parties, and interest groups depends in large part on the degree to which government replaces market or market replaces government'. Certainly, this is a primary caution which should not be disregarded when trying to study public policies on S&T. However, I think in one sense it is incomplete, and in another it could lead to some confusion. First of all, the simple dichotomy of government and market does not account for the wide and complex space of social articulations and regulations which link state action to the micro-decisions of the economic agents, thus blurring the existence of a meso-social area of quite considerable significance. Second, Lindblom's thesis hinges upon a separation between politics and the economy which is still, in part, an offshoot of the classical liberal vision. Besides, this vision becomes increasingly less productive in the face of the complex redefinitions of the boundaries between the public and the private domains which we are witnessing today.

That is why I have advocated a distinction between what I consider the two central components of a political system: the *social regime of accumulation* (SRA) and the *political regime of government* (PRG) (Nun 1987, 1989). The latter refers to the more familiar ways of conceptualizing how a particular form of state combines with a specific configuration of the political scene (in a restricted sense). It includes both 'secondary questions' mentioned by Lindblom and the pressing problems of political representation, culture and behaviour. The *social regime of accumulation* calls for a few brief comments aimed at explaining why, in my opinion, any analysis of S&T policies – in this case those of Argentina – refers the analyst to the complex trajectory of not one but at least two major processes of transition and consolidation or, if you prefer, of continuity and change.

1 THE SOCIAL REGIME OF ACCUMULATION

With this notion, I wish to designate the complex and historically situated set of institutions and practices that condition the process of capital accumulation, with the latter perceived as a microeconomic activity generating profit and decision-making investments.[2] Although the set encompasses such activity, which 'cannot be conducted in a vacuum nor in the midst of chaos', it is largely external to it (Gordon *et al.* 1982: 23). In other words, an established SRA is supported by institutional frameworks, practices and interpretations of various kinds which assure the economic agents certain minimum levels of coherence in the environment in which they operate.

The composition, features and scope of this regime vary historically from place to place.[3] However, contrary to any functionalist or reproductionist interpretation, an SRA is always heterogeneous and riddled with contradictions which elicit variable degrees of conflict. This continually highlights the vital articulatory role played by politics and ideology. Such a regime can therefore be seen as a matrix of changing configuration in which are interwoven specific strategies of accumulation and various tactics for their implementation. Thus, the accumulation of capital becomes 'the contingent result of a dialectic of structure and strategies' (Jessop 1983: 98). Consequently, the study of it goes beyond the strictly economic sphere and requires an engagement in what has aptly been termed a *political sociology of political economy* (Gourevitch 1986: 19).

It follows, then, that an SRA is a multidimensional historical phenomenon of medium or long-term, in which three major moments can be analytically distinguished: the emergence; the consolidation and expansion; and finally the depletion and decline, which may or may not lead to a generalized crisis (Block 1986: 182). Note that the latter would be a crisis *of* the regime, and should not be confused with crises *in* the regime, which normally punctuate its course without questioning its parameters.[4]

As an SRA stabilizes and settles down, the inertial force of its characteristic modes of institutionalization and practices increases, without cancelling the problematic nature of the ever present 'dialectic of structures and strategies'. What happens is a process of naturalization in which a particular organization – either of the market, or of the relationship between capital and labour, or of the connection between the state apparatuses and the corporations – may enter the prevailing common sense of the economic agents. When this occurs, any attempt to modify the status quo can be presented ideologically by the defenders of the latter as political interference, and not as what it really is: a movement of pieces on the political chessboard, on which, in principle, all parties are players.

From this point of view, it is as reductionist to imagine, for example, that among the Latin American countries a purely economic era has come to an end during the last few decades and – depending on the location – another has started, as it is misleading to phrase the current dilemma in terms of a simple opposition between state action and the market. With regard to the former, the demise of the so-called model of industrialization by import substitution has not been an exclusively economic phenomenon nor has its meaning been univocal: in some places it was the result of the decline and in others of the crisis of an SRA which had sociopolitical features peculiar to each country. Therefore, neither its nature nor its dynamics nor its present horizon of possibilities can be construed a priori as being similar.[5]

With regard to the latter, the widely discussed contrast between state action and the market ignores at least two phenonema. First, that every kind of market structuring requires state action.[6] And second, as I mentioned earlier, between those two domains extends a huge area of social regulations without which it would be literally impossible for the overall scheme to work. As Block (1990: 42) points out, the symmetrical error made by the theory of market self-regulation and the theory of planning lies in assuming that just one of these levels is sufficient to produce rational and efficient results, when 'the reality is that the efficiency of a particular economy will depend on how the three fit together (market, social regulations and governmental policies)'.

I should mention at least two important corollaries of the approach I have just outlined. One is that, contrary to what is usually done, the first moment in the definition of social classes should refer not to the economic plane (the famous position in the productive process) but to the SRA plane, i.e. to a very diverse institutional formation, stemming from a particular history, which gives prominence to certain leading actors with their own practices and ways of organization. In this context, different forms of opposition and struggle acquire their meaning, which may be reinforced or inhibited by other networks of social interaction. Thus, as opposed to economistic narrations tending to emphasize long and continuous histories of the social classes, an analysis in SRA terms points rather to the discontinuities, to the changes that take place both in the situation and the composition of the actors and in the prevailing images which shape their demands and which spotlight, each time, the concrete activities of specific groups (Gordon *et al.* 1982: 40). Some years ago, Sartre reminded the Marxists that workers are not born the day they enter the factory; it need hardly be said that neither are the bosses.

The other corollary relates to the changes which take place in the nature and the logic of the political system.[7] These changes may be due mainly to the transformations occurring either in the SRA, or in the political regime of government (PRG), or in both together. The first case is illustrated both by the transition to the 'Progressive Era' and by the second 'New Deal' in the United States. The democratization of the liberal PRG in England at the end of the nineteenth century within the framework of a well-established SRA, or the 1916 elections in Argentina, are good illustrations of the second case. As regards the third case, few examples can be as dramatic as those currently being offered by the countries of central and eastern Europe.

As I have stated elsewhere, the changes which the Argentine political system has undergone in recent times are a sub-type of the last category. On the one hand, although it is hard to argue that it was totally exhausted, by the end of the 1960s the SRA that had emerged after the Great Depression and expanded in the post-Second World War period, was coming to the end of the road in some areas. Also, at that time efforts began to shape an alternative regime, whose stage of emergence has lasted up to the present day. On the other hand, after eight years of military dictatorship, constitutional rule was restored in Argentina in 1983, inaugurating a new PRG. It is this complex combination of two transitions – towards that SRA and this PRG – and not simply the passage to democratic liberalism, which I believe gives meaning to some of the most significant political events of the last few years. Just as before, frequent changes in the PRG influenced the course of an SRA, which nevertheless managed to maintain its overall profile with considerable stability.[8]

2 INTERACTION BETWEEN THE SOCIAL REGIME OF ACCUMULATION, THE POLITICAL REGIME OF GOVERNMENT AND S&T POLICIES

Although the distinction I draw between the SRA and the PRG is essentially an analytical one (and should be understood as such), there are concrete situations where its empirical correlates appear more clearly than in others.[9]

In the ideal type of the so-called 'developmental state', for example, the distinction between the SRA and the PRG seems to disappear because the same actors tend to run both regimes, extracting surplus while at the same time supplying collective goods (Evans 1990: 10). In general these are cases where a change in the PRG is quickly followed by a change in the SRA – as demonstrated by the military coup led by Park Chung Hee in South Korea, with its immediate violent attack on the business *establishment*, many of whose members were paraded in derision through the streets. However, as soon as an emerging new SRA stabilizes, its own logic usually begins to assert itself, links between the state and the process of accumulation multiply and become more complex, and business leaders and their organizations play an increasingly prominent role – as is happening today in South Korea itself, where an enormous concentration of power lies with the major conglomerates (*chaebol*).

It is in those cases where effective 'neocorporatist pacts' have materialized that the distinction to which I refer tends to assume some of its more characteristic forms (e.g. Austria, Belgium, Holland, Norway, Sweden, etc.). This is due to the existence of explicit operating rules and relationships between a specific group of state agencies and the top organizations which represent the business community and the workers; and also to the considerable care taken to ensure that such pacts, based on a functional representation of interests, will not affect the institutional framework of democratic liberalism, which is based on territorial representation and gives support to the legitimacy of the whole political system.

In outline, then, the perspective I am presenting leads to a concrete examination of not only the ways in which particular SRAs and PRGs are structured 'internally' in each case, but also how they relate to one another. As a first approach, this may be visualized as a continuum, ranging from a high to a low relative degree of differentiation, specialization and separation between its respective components. However, certain observations need to be made.

First, it is important to stress the word 'relative': a high degree of differentiation, as in the liberal democracies, does not mean that interest groups, often supported by sympathetic sectors of the state apparatus, would not keep up pressure on the political parties, the parliament or the executive to obtain the laws and regulations they need. Even when the degree is low, as in the recent military dictatorships in Latin America or in the more extreme experiences of the so-called totalitarian regimes, there are always backroom talks and 'trade-offs' between those promoting the process of accumulation and those trying to ensure compliance, giving rise to various forms of that 'black parliamentarianism' which Gramsci once labelled as characteristic of Italian fascism.[10]

The second point refers to the kinds of links I am speaking about and recalls the criticism which Bachrach and Baratz (1970) once addressed to the studies of pluralist orientation on urban politics in the United States. These authors introduced the idea of '*non-decision-making*', precisely to demonstrate that besides their explicit ways of influencing the decision-makers, the power nuclei are also the ones who define the agenda and decide which questions should *not* appear in it, setting limits to what is or is not decidable. Such '*non-decisions*' provide a warning of the presence of political and ideological struggles which are usually unleashed in any attempt to enlarge, for example, the field of scientific and technological policies, and highlights the recurring naïvety of those who believe that an administrative resolution is all that is needed.

The third *caveat* concerns the excessive level of generality encountered when speaking in terms of a continuum. This approach is really more useful as an outline description of the subject than as a framework for the study of specific public policies. The latter certainly calls for prior examination of the particular configuration of the policy area concerned, and it would be reductionist to assign a priori to this area those characteristics that can be attributed to the overall system. It is relevant here to mention a qualification which was formulated with respect to another continuum, similar to the one I am discussing and which is also pertinent to my argument: the continuum by which states are classified from weak to strong according to the power they hold in relation to their own societies. As Zysman rightly observed with regard to this second continuum:

> a government's ability to act in one policy area will be quite different from its ability to act in another. Thus, France is a strong state in terms of energy policy, but in the social services the bureaucracy is trapped in a morass; in welfare policy, France is a weak state. The concept of a strong state–weak state continuum refers to a generalized capacity, not to specific abilities to carry out particular tasks. The policy tasks in each sector vary, as does the pattern of interest organization.
>
> (Zysman 1983: 297)

Finally, expanding on Vogel's analysis (1986) of regulatory practices, one should not lose sight of certain 'national styles of policy-making' which transcend and permeate the relations between particular SRAs and PRGs. As Vogel maintains:

> There is thus no necessary relationship between the organization of a nation's political economy and its approach to government regulation: Britain and the United States are roughly similar with respect to the former but very different with respect to the latter.
>
> (Vogel 1986: 268)

This is certainly a theme that should be included in analyses of public policies on S&T, as is already being fruitfully done in other areas.[11]

3 ARGENTINA, CANADA AND TECHNICAL CHANGE

At this point I believe a comparative example may be useful to clarify the discussion and at the same time show how it can contribute to a better interpretation of the processes of technical change. In literature, attention has been drawn to the similarities between agricultural developments in Argentina and Canada. At the end of the last century, both countries were characterized by an abundance of land and a shortage of labour, by their fertile soils and by their rapid and enthusiastic welcoming of foreign workers and foreign capital. If there was any natural advantage it certainly fell to Argentina: first, its climate was much milder and temperate throughout the whole year; and second, because of its geographical location the Argentine Pampas was close to the sea ports, a facility not available to the Canadian prairies.

At the beginning of this century, both countries became not only two of the world's largest wheat exporters, but also the grain came to be their main export. The similarity was all the more in that the two nations sent this produce to the same foreign markets, and, except during the First World War, received the exact world price. Yet, whereas until 1910 Argentina produced and exported more wheat than Canada, afterwards the situation was reversed. Furthermore, from then onwards 'the Canadian prairies were not only outpacing the Argentine Pampas in quantity, but also surpassing them in the quality of wheat production' (Solberg 1987: 2). In fact, Canadian wheat gained a reputation on the international market which the Argentine product never achieved. Moreover, this trend was maintained for several decades, during which Canadian grains became clearly dominant in the international market, whilst Argentine agriculture went into a stage of open decline.

What happened? I believe the situation is difficult to understand without first comparing certain aspects of the SRAs and PRGs which were emerging in the two areas, and the bearing they had, in this case, on the respective wheat production techniques.[12] The SRA which was taking shape in Canada in the latter part of the nineteenth century, was grounded on a comparatively strong state which, led by the Conservative Party between 1867 and 1873, and again between 1878 and 1896, implemented the National Policy of Sir John A. Macdonald, political spokesman for an economic elite composed mainly of traders and bankers. Since the middle of the century, this eastern Canadian elite had started to invest in industry and had become a great defender of protectionism. At the same time, it soon realized the importance of expanding westwards, both to check the territorial ambitions of the United States and to join the world market as an exporter of grain.[13]

Hence, the three fundamental components of the National Policy were:

1 the tariff, to protect and promote industrialization;
2 the building of a transcontinental railway to link the east with the west of the country;
3 a massive immigration programme, to populate the prairies so that they could become a source of primary exports and a market for the protected industries in the east.

The coherence of this great plan for national development headed by the state is evident. It established the framework of the SRA which was to prevail in Canada until the 1930s, and to which the Liberal Party also adapted its actions when it came to power.[14]

There was no such design in the case of Argentina. To begin with, unlike the situation not only in Canada, but also in the United States and Australia, the economically hegemonic sectors were not the traders and the bankers but the landowners of the Pampas, whose main activity was cattle ranching. Then, the SRA which the so-called 'generation of 1880' succeeded in articulating was imbued with a free-market ideology which became part of the national common sense, further confirmed by the great boom derived from the incorporation of the country to the international division of labour. From this viewpoint, it made good business sense that the railways, for example, should be in the hands of foreign capital, since their construction increased land values and provided an outlet for produce without requiring local investments – although it did very little to integrate the nation effectively. It was also considered quite natural and sound that, with the backing of the theorem of comparative advantages, a major portion of the foreign exchange earned from exports should be allocated to importing most of the industrial goods needed, instead of manufacturing them locally. In this context, the role assigned to the state was clearly a subsidiary one. The dictates of the market and not public plans or policies were to be the main engine of progress. In the words of Julio A. Roca, twice President, the government's role was simply to provide 'peace and administration'.

Certainly, during the emergence of this particular export-oriented SRA, it was also essential for the state to encourage the arrival of immigrants and the laying of tracks. But, I insist, this did not go much beyond the role of promotion and did not become a constituent part of a true official development programme, in the style of the National Policy.

At this point the relationship between the respective SRAs and PRGs becomes particularly relevant. In Canada, the cornerstone of the colonization of the prairies was the Homestead Act of 1872, which granted plots of public land to anyone wanting to settle. Contrary to what occurred in Argentina, based on this law the immigration policy did not stop at attracting foreigners but set out to install them permanently as farmers on condition that they became naturalized. This process was reinforced by the fact that 'one of the two major parties – the Liberals – attempted to build a political machine on the basis of the immigrant vote' (Solberg 1987: 29). The situation in Argentina was quite different. There was no public land on the Pampas to share out, so many of the new arrivals settled for renting plots offered by the ranchers, for three or four years, with the express condition that they devote themselves to agriculture – in order to improve the soil – and not to cattle rearing.[15] This, plus the fact that the immigrants were mostly Italians who tended to return to their home country, a phenomenon which occurred everywhere and not just in Argentina, contributed to the marked 'swallow' character of immigration (a feature of more than 50 per cent of the total flow between 1890 and 1914).[16] Furthermore, the two major parties, the Autonomists and the Radicals, did not wish these foreigners to become citizens since they were viewed as a potential threat in the voting booth.[17] The figures speak for themselves. According to the 1911 census in Canada, 47 per cent of the immigrants were already naturalized, whilst in Argentina the 1914 census revealed only a 1.4 per cent.

It is in this context that light is shed on a series of contrasting features in wheat production in the two countries – differences which are of immediate relevance to the subject of technical change. Most important, production in one case was in the hands of small landowners, and its development depended increasingly on intense mechanization to extend the cultivated area and the productivity of family labour; while in the other case, the main concern – particularly of the temporary tenants – was to increase the productivity of the worker rather than the land, with a minimum possible investment in machinery.[18] At the same time, in Canada, bank credit was readily available to farmers, while the latter, as citizens, were gaining increasing political strength – first via the Liberal Party, and later via the Progressive Party – to defend themselves against the abuses of the financiers and the railways and to promote their own demands. For the reasons already stated, this did not occur in

Argentina, where credit went to the privileged cattle-rearing landowners and not to the small grain producers, who lacked the requisite political voice to make their aspirations heard.[19]

Last but not least, consistent with my previous comments on its central role in the SRA, the Canadian government took an early interest in applying technical changes to agricultural production. Thus, as far back as 1886, the Ministry of Agriculture initiated major research into vegetable genetics on the prairies. This soon created a very favourable climate for innovations in the rural areas, which enjoyed the support of the authorities, the newspapers, the agronomists and, of course, the dealers of farm machinery. As Adelman concludes (1991: 26): 'Farmers were encouraged from all sides to make their farm the epitome of scientific operations.' This was not the case in Argentina, where it was only in 1898 that a Ministry of Agriculture was established, which took until 1912 to initiate a sporadic effort in wheat seed genetic improvement. If there was any point of relevance here to agriculture it was 'a notable official absence in technological matters' (Barsky 1988: 72). This author points out how this lack of long-term state policy in the sphere of technological generation, the slow improvement of the transport system, and finally the weakness of the agricultural machinery supply industry itself, combined to hinder technical change in grain production (Barsky 1988: 76). Added to this, as I have maintained, is the decisive role played by the different class configurations and political systems which evolved in the two countries.

Considering the foregoing, it is hardly surprising that according to official data, around 1914, agricultural producers in Saskatchewan were investing 4.3 times more per production unit in machinery alone, and 6.2 times more per hectare, than their counterparts in the province of Buenos Aires (Adelman 1991: 49).

4 THE PLACE OF THE S&T COMPLEX IN ARGENTINA

What occurred in wheat production is an illustration of the secondary role occupied by the autonomous development of science and technology in the export-oriented SRA, based on primary goods, which emerged in Argentina at the end of the last century and reached its peak in the 1920s.[20]

As is well known, in its time this SRA, which availed itself of the natural endowments of the Pampas, made the country one of the richest in the world, and certainly the most modern in Latin America. This led to a flourishing cultural life and the growth of the universities, with the university diploma seen as one of the best credentials for ascending the social ladder in a relatively open society. The other side of this evolution was that the universities became basically places for the circulation and not the generation of knowledge, while its faculties were conceived almost exclusively as schools for training professionals.

As the first Argentine Nobel Prizewinner, Bernardo Houssay, was to say in 1922: 'improbable as it may seem, most of the men at our university do not understand the role of research'.[21] Besides, given the state monopoly of higher education and the absence of official S&T policies, such professional orientation left very deep marks and this was institutionally reinforced by the meagre resources allocated to research and by the lack of full-time professors. Since then, many circles have been penetrated by 'an image of tertiary teaching as a subsidiary activity, an appendage of other more important and serious occupations' (Myers 1992: 93).

There were, of course, some important exceptions. These were particularly evident in the fields of medicine (whose model was the Institute of Physiology of the University of Buenos Aires, directed by Houssay himself); chemistry; physics and mathematics. As regards the link between research and production activities, it was extremely tenuous, concentrating basically on a few farming and mining projects (particularly geological explorations relating to oil).

The conclusion is quite clear therefore: *in Argentina, the SRA based on agricultural exports tended neither to encourage nor to incorporate the generation of scientific and technological knowledge as a central component, but to metabolize that obtainable elsewhere whenever necessary and/or profitable.* Phrased in the terms I suggested earlier, the meagre efforts which did take place (almost always in response to initiatives by the public sector) were inscribed rather *in the space of the Political Regime of Government than in the space of the Social Regime of Accumulation.*[22] This probably helps to explain a double tradition among Argentine researchers: on the one hand, their reluctance to give special importance to the possible specific applications of their findings, and on the other, their well-known political sensitivity and engagement. These tendencies, clearly intensified by the instability and the increasingly frequent changes of a PRG which could affect them greatly, encouraged a defensive withdrawal which further deepened the gulf separating the SRA from the incipient achievements in S&T.

Many things began to change in the country in the 1930s, with the advent of a new SRA, supported this time by active state intervention. If the military government of General Agustín P. Justo laid some of the foundations of such transformation, it was with the ascent to power of Peronism that it acquired its specific identity. Thus, an SRA oriented to the internal market, highly protected, and driven by the public sector was consolidated. It was to characterize Argentina until the second half of the 1970s. For different reasons, the two major social classes that developed during this period shared a high degree of heteronomy; and it was this heteronomy which became one of the dominant features of the subsequent industrialization process of import substitution.

The Peronist variant of populism promoted the development of a unionized labour movement, strongly state oriented and whose hegemonic nucleus was composed of the industrial unions. Apart from a few notable exceptions, it was a highly bureaucratic and centralized force with scant display of internal democracy and a markedly corporative ideological bent. Its constant invocation of 'social justice' almost never went beyond the lukewarm reformism of its initial themes of the 1940s: the defence of wages and a vertically structured labour organization; the strengthening of the internal market; an integrating nationalism; and a vocal opposition to both 'oligarchic liberalism' and 'unpatriotic classism', in the name of an unwavering adhesion to Perón's leadership and his project of an 'organized community' in which the representatives of capital and labour could coexist harmoniously under the tutelage of the government. It was, in sum, a unionism essentially adept at pressing for short-term, pragmatic demands, which felt less comfortable taking up government responsibilities (as happened between 1973 and 1975) than acting as the opposition in the negotiated defence of its sectoral interests (Torre 1983: 147). From this perspective – and unlike what happened in other labour movements, such as the Italian for example, which drew on the nineteenth-century tradition of scientific progress – the problems of science and technology never achieved prominence in its debates nor did they become of significant concern to the class as a whole, regardless of how important a professional title was for their children.

As for the industrialists, they also developed a strong orientation towards the state with a great dependence on the direction taken by public policies, particularly those concerning import tariff barriers which shielded them from external competition. But above all, their process of accumulation was largely reliant on the workings of the financial system, since their investments were clearly associated with the actual subsidies they received in the form of credits at real interest rates which, between 1945 and 1975, were permanently negative due to inflation. As one specialist has remarked, probably more profit was yielded this way than by the actual running of the business, or at least the losses suffered by the latter were offset (Arnaudo 1987: 162). Therefore, he added, the business community did not oppose the great inflationary explosions, which wiped out their debts. Apart from some exceptions, if this situation led anywhere in terms of technology, it was towards importing it and, after 1958, bringing it into the country via subsidiaries of transnational companies. In any event, the greatest (and not inconsiderable) efforts were directed to seeking ways of adapting such techniques – clearly designed for other settings – to the characteristics and size of the heavily protected local market and without any concern regarding international competitiveness.

This very brief historical outline would not be complete without mention of what happened between 1964 and 1974, during which period both the GNP and the industrial production saw high positive rates of uninterrupted growth. Indeed, the highest levels of industrial growth in the history of Argentina were achieved during this period, together with a sustained rise both in productivity and in the share of manufactured goods in total exports. Furthermore, during those years, due to the attained self-sufficiency in oil and a considerable export expansion (firstly, primary goods; later also industrial ones), the familiar and painful stop-and-go cycles of the Argentine economy came to an end and there was a noticeable reduction in the chronic instability which had afflicted it since the post-war period (O'Connell 1985: 38–9). Despite the fact that questions of industrial policy and technical change were still relegated to the public agenda, it is significant that a process of selling technology and engineering services of local origin to other countries was initiated, and that the development of more sophisticated sectors, such as electronics, pharmaceutical, chemistry and machine tools began to gather momentum. Indicators such as these led Katz to surmise that

> *pari passu* with the industrialisation by import substitution, a long term coming-of-age [was] taking place in the domestic production sector, a process which without doubt originates in the gradual development and consolidation of the internal technological capacity of our manufacturing industry.
>
> (Katz n.d.: 3)

None the less, this process was to be drastically interrupted in 1976 by the change in the PRG brought about by the military coup headed by Jorge Videla. Although the inspiration for this coup was not primarily an economic one, and moreover within the top echelons of the military there were differences concerning this matter, the neoliberal line represented by José Martínez de Hoz was finally imposed, supported by the leading sectors in finance and agriculture. What happened – the relative boom I have just mentioned notwithstanding – was that one of the pillars of the then current SRA was visibly on the verge of collapse in a country practically devoid of a capital market. I am referring to the state and its financing capacity, seriously impaired by an unstoppable inflation and the defensive reaction it brought about, i.e. a dollarization of savings which began in the 1970s and which naturally caused a serious reduction in the volume and term of the investments in national currency. Seen in the form of an allocation of resources conflict, what was really at stake was the very future of that SRA, given the features which I described earlier.[23] And what was going to take place was its liquidation on the worst possible terms.[24]

Viewed from the angle that concerns us here, I shall in due course refer to the transformations that took place in the agricultural sector. As for the industrial one, it entered a phase of contraction and regressive restructuring and 'the processes of industrial and technological ripening which had developed during the previous decades were interrupted in key sectors for technical progress and productive integration, such as the metallurgical industry, electronics and the capital goods industry in general' (Nochteff 1991: 342).[25] Against this background, not only did science and technology issues disappear from the public agenda but, as had occurred before in 1966, both the universities and specialized state institutions and their teaching and research staffs were subjected to a brutal 'witch hunt' which decimated their numbers and broke up painstakingly formed teams.

In 1983, there was a further PRG change: the military relinquished power and a constitutional government took over, headed by Raúl Alfonsín. The legacy the new authorities received from the dictatorship was a country in ruins, with a per capita product lower than in the preceding decade, the fiscal coffers empty, a spectacular fall in the rate of investment, a functional distribution of income at least as regressive as that prevailing in the era when Perón first achieved fame, a three-figure annual inflation rate which no other nation had withstood for such a long period, and to crown it all a crushing external debt. In one sense, the decline of the SRA which had emerged during the 1930s was consummated. Around 1950, when this SRA established itself, the product per capita in Argentina was higher than in Japan, Italy, Spain, Austria or South Africa; in 1983 it had fallen significantly behind all of these countries.

At the same time, however, in a contradictory and non-articulated way, the outline of a new SRA was emerging, and with it a fierce struggle among economic groups to defend and/or improve on the positions they had acquired or were in the process of acquiring. And it is exactly this point whose real dimension the newly installed authorities were to understand only later: the 'dialectic of structures and strategies' is never more ruthless than during the emergence of an SRA, when the ground won is still so precarious and the perceived horizons still so uncertain, particularly in today's increasingly international scenarios. In addition, in order to clear the way for the new SRA it was necessary to redefine tangled regulatory networks and create new ones, with the consequent conflicts of interests and institutional readjustments that this always generates.

Thus, the other side to the deindustrialization of the preceding years and the downgrading of those activities which result in technical progress, was a busy process of acquisition and merger which led to a growing concentration of markets without a real increase in installed capacity. Above all, a group of companies and conglomerates of national origin consolidated and together with certain transnational conglomerates, ended up as the main beneficiaries of the regime of force which devastated Argentina. This outcome (partially related to the introduction of technical innovations only, in what Nochteff 1991 calls 'greenhouse enclaves') was encouraged by the biased orientation given to the purchasing power of the government and the public corporations; the way in which privatized activities and companies were awarded; and, to top it all, the policies of industrial promotion, which replaced the previous credits at negative interest rates as a suitable mechanism for capital formation almost entirely at the expense of the exchequer, that is to say, the population as a whole.[26]

It is apparent, therefore, that the potential emergence of a new SRA can trigger the simultaneous creation or strengthening of some of its main economic agents, which may or may not already be in existence when the process begins. This does not mean that there is a predetermined outcome in each case, and it is also far from the elimination of recurring ideological and political conflicts. However, in order to influence the outcome, a fairly clear diagnosis of the situation and a significant degree of power are needed. The Radical Party came to government lacking such a diagnosis and was unable to tap the considerable amount of popular support which it enjoyed at the beginning, in order to guide the restructuring that was taking place along the moderately social democratic lines its discourse suggested.

After a short while, and contrary to its initial convictions, the Alfonsín government abandoned popular mobilization and protagonism as the basic resources of its political action. By doing so it forgot that in present-day Argentina the government itself has to compete for power, even more so when its aim is to alter the status quo. For this purpose it was essential for the authorities to enjoy wide, organized and active public support, as the President himself clearly maintained in his inaugural speech. The justification which was to be repeated *ad nauseam* was voiced quite early by a Radical Party MP: 'The Alfonsín government is the government of the possible'.

However, support was becoming more and more necessary in the face of intense pressure from the United States which, at the head of Argentina's creditor countries, significantly limited the degrees of freedom of 'the possible'. It will be recalled that when, in February 1985, James A. Baker III replaced Donald Reagan as Treasury Secretary, there was a marked shift in US external policy on economic matters. Until then, with regards to debtor nations, its essential features had been the strengthening of the IMF and austerity requirements (trimming the public sector; adjustment of the economy to encourage exports to obtain foreign currency needed to pay off international debts; etc.). With the prospect of a more relaxed attitude towards debt, it added the requirement that 'structural reform' programmes should be implemented, the parameters of which were to be the liberalization of foreign trade, the privatization of public companies, the opening-up to foreign capital, and the deregulation of business activity.[27] Gradually and hesitatingly, but increasingly dependent on Baker's support, the Argentine government gave in to these demands, thus steering the emerging SRA in a particular direction and making a national development strategy unviable in practice.

In July 1987, the euphoria generated by the 'Plan Austral' already spent, and faced with a new inflationary spiral and a dramatic fall in the trading surplus, the then Minister of the Economy, Juan Sourrouille, declared: 'What we Argentines are experiencing . . . is the crisis of a populist and simplistic model, a closed model, in short, a centralized and state oriented model.' The problem, more political than economic, was how to replace this model and which coalition of forces should be assembled to handle the task. This was owing to the fact that although it was true that 'the Argentines were experiencing' a crisis, it was also true that they were witnessing a cut-throat competition of interests among various local and foreign economic groups which were trying to solve it to their own advantage.

No doubt, the task was as arduous as it was complex, and the government showed it was not capable of tackling it successfully.[28] Increasingly deprived of a solid political base of its own, the inconsistencies and tensions which arose from its overly unstable and conjunctural alliances were reflected from the third quarter of that very year, 1987, with a fall in the level of economic activity: a decline in industrial occupation, in hours worked and in the average real wage. Even worse, the rate of inflation more than doubled that of the preceding year, once again reaching the dreaded three figures. Neither were the vagaries of the PRG attuned to the emerging definite course of the developing SRA, nor was the economic team able at that point to promote anything other than short-term solutions. Less than two years later, in the midst of hyperinflation and mob ransacking of supermarkets, Alfonsín was forced to resign and hand over the government prematurely to Carlos Menem, the Peronist candidate who had just won the presidential elections.

Despite several initial comings and goings which the 'Cavallo plan' put a stop to in 1991, what followed can be seen as a resolute intensification of the concentrating, exclusionary and liberalizing course which the new SRA had started to adopt during the last military dictatorship and which the Alfonsín government did not reverse.[29] Until now, the greatest achievements of that plan have undoubtedly been monetary stability and economic reactivation, although since the end of 1992 the latter has been losing momentum. At the same time, an uncontrolled process of privatizations – some of which have been rather disastrous for the country – has caused the state apparatus to shrink, while allowing the major economic groups to do very good business indeed (according to certain estimates, by the time this process will finish 50 per cent of the Argentine GNP will be in the hands of a dozen or so companies). As is well known, the size of a state is not necessarily related to its strength. In this case we are seeing the transition from what had become a large and weak state to a smaller one in open and deliberate retreat, which so far has been increasingly removed from development concerns and from the active promotion of genuine scientific or technological policies.

In a context similar to the one I have just outlined – with admittedly heavy and selective strokes – it is not surprising that generally the literature tends to agree with an assessment made recently drawing attention to

> the lack of resources suffered by the Scientific-Technological Complex and the University; the poor management; the lack of internal articulation; the absence of links with the production sector; the non-existence of a medium and long term strategy in science and technology as a component of a national development strategy; the lack both of an industrial strategy [and] a suitable human resources policy in Argentina.
> (Oteiza *et al.* 1992: 15)

As in other parts of Latin America, attempts at organizing and co-ordinating this complex at the national level began at the end of the 1950s, by transplanting institutional models which were being introduced at the time in Western Europe (particularly France) and which gave the state a leading role in these matters.[30] This meant increasing the interest in technology which had led the first Peronist government to establish a Ministry of Technical Affairs in 1947, by also making room for basic scientific research in the official plans. Thus, the National Committee for Scientific and Technical Research created in 1958, aimed at 'co-ordinating and promoting scientific research' to the extent that such activities 'promote the improvement of public health; a wider and more effective use of natural resources; an increase in industrial and agricultural productivity; and, in general terms, the welfare of the population as a whole' (Decree-Law 1291/58). The point I want to make is that, put in the terms I have been using and consistent with what I stated previously, this creation clearly issued from the space of the PRG and not the SRA. To quote the words of Caldelari *et al.*:

> The creation of CONICET was more a response to demands from the scientific community; the state's need for prestige derived from its support of scientific development; and the desire to modernize – fashion of the time – *than to an effective demand from the sectors of production.*
> (Caldelari and Casalet 1992: 171, my emphasis)

Predictably, this demand did not grow subsequently either. Furthermore, although CONICET has tried hard since 1984 to encourage this growth, the results so far have been rather poor.[31]

Also within the framework of the PRG was the creation in 1984 of the Department of Science and Technology (SECYT). At first it was subject to the Ministry of Education and Justice and, since 1989, to the National Presidency, which also founded and placed under the jurisdiction of the SECYT the Federal Board of Science and Technology, set up to integrate national and provincial initiatives in the field of S&T. The tasks of the SECYT are to advise the President in all matters relating to S&T; to design policies of scientific and technological development; and to promote research, financing and transfer of know-how in these areas. To this end one of the main tools used by the Department is the national programmes in various specialities: biotechnology, petrochemicals, food technology, non-conventional energy, endemic diseases, etc.

Once again, the effectiveness of this department is being seriously impaired by a recurrent lack of adjustment between its objectives (always ambitious) and its resources (always scarce) but especially by its difficulty – which can no longer be considered merely conjunctural – to relate to the SRA. For this reason, notes Oteiza (*et al.* 1992: 27), even when the Executive formally approves projects submitted to it by the SECYT, 'the force of an economic policy running along a different track imposes a different logic and marginalizes the S&T Complex within the government itself'.[32]

Although the SECYT and CONICET, together with the Board of National University Rectors, are the highest co-ordinating organs of this complex, the disparity of their resources is glaring. In 1988 for example, the SECYT received less than 5 per cent of the itemized allocation of funds for S&T on the national budget, while the proportion allocated to CONICET was over 40 per cent. It is also significant that only three institutions – CONICET, the National Atomic Energy Commission (CNEA) and the National Institute of Agrarian

Technology (INTA) – absorb most of the public funds assigned to S&T: in that same year, 1988, in aggregate terms, their share of the budget amounted to nearly 80 per cent (Azpiazu 1992: 199).

The above fact is of additional interest here because the last two bodies mentioned are almost unanimously recognized by students of the subject as examples of successful activity in the sphere of science and technology in the country. This leads one to ask what the main reasons were for this success and to what extent, if any, they constitute exceptions *vis-à-vis* my central argument.

5 THE CASE OF INTA

It is interesting to note an observation that coincides with our present discourse:

> If there is a single element common to the whole productive process on the Pampas until the end of the Fifties, it is the scant interest shown by the various sectors which succeeded one another in political power to build a state structure suitable for developing and disseminating farming technology.
>
> (Barsky 1988: 81)

This 'scant interest' ran parallel to the sharp decline in grain production in the 1940s and 1950s, which the modest expansion in stockbreeding at the time was quite unable to compensate. It was only in the 1960s that the maximum levels of agricultural production attained in the 1930s were restored (and surpassed since 1965). Statistics give clear testimony of the magnitude of the transformations that took place: between 1962 and 1984, for example, the value of agricultural production on the Pampas tripled; the productivity of the land more than doubled; and labour productivity quadrupled (Obschatko 1988: 117). The main point stressed here is that about 80 per cent of this expansion was a result of the technological changes that were introduced, whilst the other 20 per cent was due to the transfer to agriculture of land previously used for cattle raising.[33]

A particularly prominent role in those technological changes was played by INTA, founded in 1956 on the express recommendation of Raúl Prebisch. If this recommendation is alluded to it is because it was in tune with the role generally assigned by ECLA to the state as the agent for transforming society – a then dominant view which also helped to induce other Latin American governments to participate more actively in the development of agricultural technology. There were other contributing factors, such as the actual decline of rural production and the legitimacy which the Punta del Este Conference and the international assistance programmes that followed gave to that participation. What should be noted, however, is that INTA did not come into being as a result of the pressure exercised by the producers or their associations: *it emerged within the framework of the PRG and its merit lies in having then implanted itself successfully in the SRA*. It is worth asking how this was possible.

There is, first of all, a well-known peculiarity of many of the innovations required by rural enterprises: they cannot be patented or monopolized by those who introduce them, so that a good deal of the research efforts involved by the development of new techniques usually tends to be 'external' to these enterprises (Sábato 1981: 64).[34] This resort to external sourcing is heightened by the fact that the technological policies for this sector – unlike price or loan policies, which are usually fairly product specific – require an overall strategy which only governments are in a position to build, given suitable conditions, indeed had been the case since the mid-1950s (Piñeiro and Trigo 1983: 237).

Besides, apart from some misgivings initially voiced by the Rural Society, while the producers' associations did not push for the creation of INTA, neither did they try to prevent it. The previously mentioned climate of opinion and the critical stagnation suffered by agricultural production were no doubt relevant factors in this connection. Afterwards, when INTA had been consolidated and the agriculturalization process relaunched, further reasons appeared. Thus, in the 1970s, world trade in the type of grain cultivated in the Pampas doubled; and since world production itself grew by less than a third, it follows that most of the growth was concentrated in the exporting countries (CGE 1990: 222). This means that competition among the latter intensified and it was vital for Argentina not to lose markets or its position in them.

The work of INTA would not have borne fruit, however, had it not been successful in including itself in the sectoral logic of the then prevailing SRA. There were at least three factors which favoured this inclusion. The first is that INTA was not seeking to redefine but rather to attach itself to the economic model which traditionally ruled agriculture in the Pampas. It should be emphasized that INTA is statutorily barred from exercising inspection or control functions in farming production. Its two purposes are research and technological transfer only, and these have remained unchanged since the beginning. This is quite atypical when compared with the evolution of similar bodies in other countries.[35] The second factor concerns the network of regional centres, experimental stations and extension agencies through which INTA has been primarily concerned with promoting direct and permanent links with the producers and their problems.[36]

There is a third, more encompassing factor which was fundamental in making possible what has been termed the Argentine 'agricultural revolution'. On this point it is worth quoting Sábato:

> the decisive element originated in initiatives taken outside agriculture, especially by the governments and state institutions, which modified the conditions of demand and supply of technical innovations and whose adoption raised agriculture on the Pampas to a higher level of productivity. More to the point, we maintain that the changes in the demand for technology were induced by the creation and continuity of a policy of credits and tax allowances which meant subsidies for capital investment, whilst the technological supply side was promoted mainly through the creation of institutions, particularly the INTA.
>
> (Sábato 1981: 86)

Of course, this does not detract from the commendable work carried out by an institution which has been noted in general for its good leadership, its technical capacity, its effectiveness and its ability to adapt to new situations. But its achievements would have been quite different had it not enjoyed firm state support at the level of the SRA and in accordance with the latter's dominant orientations. Moreover as stressed by Lattuada (1988: 107–11), the only agrarian policy which remained constant over a long period, despite the frequent changes of government and even a PRG, was the transfer of income to bring technology to the countryside via subsidized loans and tax exemptions.

Sustained state action, then; yet also a style of intervention which strongly favoured the private process of capital accumulation and did not upset vested interests in the sector. Years later, when the Alfonsín government tried to introduce a more extensive and systematic rearrangement of rural activities to speed up the incorporation of new technologies and stabilize the internal price of agricultural products, determined opposition by the 'anti-interventionist' big agrarian corporations caused the abandonment of the project. Without substantial modification of the balance of power, reform of the sectoral SRA clearly belonged to the field of 'non-decisions' (Nun and Lattuada 1991: 21–72).

This point is well illustrated by contrasting the experience of INTA with that of the National Institute of Industrial Technology (INTI). Although created in the same period as INTA, the influence of INTI in the technological overhaul of industry has never been significant (Oteiza *et al.* 1992: 43; Valeiras 1992: 149–60). No doubt part of the reason lies in the difference between the nature of this sector and that of agriculture (for example, the entrepreneurial possibilities of obtaining technological quasi-rents). *But what I see as a more significant decisive factor is the great difficulty INTI has had in integrating into the sectoral SRA owing to the absence of consistent and sustained industrial policies, which could have favoured that integration.* It has therefore acted basically as a provider of services (quality control, metrology, chemical analyses and material testing, particularly for the automotive industry) and not as a source of technological innovations – as shown, for instance, by its lack of participation in the reconversion process of the electronics industry which began in the mid-1970s. One consequence is worth highlighting: because of its weak integration in the SRA – its lack of *embeddedness*, to use Polanyi's term – INTI has remained much more exposed than INTA to all the vicissitudes of the PRG, as witnessed by its constant changes in management, particularly since 1973.[37]

6 THE CASE OF THE CNEA

The bracket into which the CNEA fits in this discussion immediately signals its unusual nature: namely, an organization that has emerged and taken shape essentially within the framework of the PRG, in spite of how it may have influenced certain aspects of the SRA.

As long ago as 1945, a decree originating from the Ministry of War established the strategic value of uranium minerals.[38] The CNEA was formed five years later, based on the premise that 'progress in research relating to atomic energy cannot be ignored by the State, in view of the many implications for the public domain that its practical application is having or may have in the future' (Decree 10.936/50). Shortly before and in the midst of great publicity, the Perón government had given wide support to a dubious nuclear project on Huemul Island, initiated by Ronald Richter, which ended in complete failure in 1952. This is when the CNEA really started operating; and in all probability the fiasco they had experienced strongly influenced the authorities in recruiting the new staff from a wide and pluralist spectrum, paying attention to the qualifications and not to the ideological leanings of the researchers (Adler 1987: 295).

Sociological literature has often emphasized that both the time when an organization comes into existence and the features that distinguish its actions from the beginning, carry importance for its future. The CNEA provides an excellent example of this, reinforced by the relative isolation and great stability which surrounded its growth. Until 1984, i.e. for more than thirty years, Navy control guaranteed this stability. At the time of the CNEA foundation, the quest for nuclear autonomy was certainly an objective dear to the military, but on which there was considerable agreement by other segments in Argentine society. The same applied when, shortly later, the goal of stimulating the development of Argentine industry (especially metallurgy) by means of contracts with local suppliers was added to the agenda.[39] Apart from the pluralistic atmosphere which characterized it from the start, the CNEA also benefited from effective and solid leadership (Admiral Quihillalt's administration lasted for eighteen years, during which time Argentina had six changes of President) and from a team of exceptional researchers who were firmly committed to the goal of promoting an independent technological and industrial development.[40]

However, my purpose here is not to review the rich history of the CNEA (including some negative aspects as well) but rather to highlight certain attributes that clearly make it a very special case.

1 Because of the strategic value of its mission, it was given a niche in the PRG where it was protected as far as possible against political vicissitudes.
2 Both its initial objectives and those added later (in the 1960s, energy generation; in the 1970s, operation of the nuclear power plants) did not elicit major opposition, at least in an open form.
3 It is only after this particular framework is taken into account that Adler (1987: 298) is right in saying that 'the first and foremost factor in the success [of CNEA] was institutional'.
4 The institution and its managers are to be credited with the effort to link it to the SRA by the privileged position granted to the national suppliers, thus promoting and assisting in the realization of true 'technological leaps' in the production lines of these suppliers.
5 It was also on the initiative of CNEA that since the mid-1970s joint ventures were started, with various juridical formats and different degrees of participation. This initiative was maintained after 1983, despite the serious financial difficulties the organization has been facing since then. These difficulties led to the halting of its largest projects, still in the making, and caused a major and regrettable exodus of top-level researchers, due to low salaries.[41]

At present, CNEA's future is rather uncertain: while its authorities wish to transform it into a public holding with some participation of private capital, other sectors of the government – in line with the neoliberal spirit of the age – are apparently planning to dismantle the central organization which was so crucial in the operation of the CNEA and to divide it into four corporations, with majority participation of private capital.[42] It is interesting to note that even a conservative newspaper like *La Nación* has warned in a recent editorial (10 June 1993) that, given the nature and functions of the

CNEA, it should not be assimilated with the other state enterprises the government is currently privatizing. At any rate, the CNEA's present problems are not just financial and concern the future of nuclear energy in Argentina, now that the huge expectations placed on it thirty years ago have dissipated.

7 THE STATE AND TECHNOLOGICAL INNOVATIONS IN ARGENTINA: POSSIBILITIES AND LIMITATIONS

The cases of INTA and the CNEA point clearly to the developmental potential Argentina has in the field of S&T, both as regards organizational and managerial capacity and the level of human resources that are available. They also show unequivocally, though, that to realize this potential, a firm and sustained commitment by the state is essential.[43] To return to Zysman, quoted some pages back, these examples reveal to us in fact *policy areas* in which the state has shown relative strength, in one case reinforcing the private process of capital accumulation and in the other case motivating by strategic considerations.

Beyond these areas, the picture becomes negative, as may be gathered from a glance at the public budget allocations.[44] It is useful first to consider what portion of GNP is allocated to S&T by other countries, such as Japan (2.9 per cent), United States (2.8 per cent) and the members of the European Community (2 per cent combined average). Plus the fact that the Academy of Sciences for the Third World, based in Trieste, has estimated that the minimum investment needed in a developing country in order to secure a critical mass in S&T is around 1 per cent of the GNP.[45] Against this background, in Argentina, between 1983 and 1988, this proportion fluctuated around 0.38 per cent, fell to 0.2 per cent in 1990, and currently does not exceed 0.3 per cent (Oteiza *et al.* 1992: 50, also *La Nación* 8 Fe. 1993). And this without taking into account the partial wastage of these funds. To this has to be added the tiny private sector share in S&T expenditure and investment, which does not normally seem to exceed 5 per cent or 6 per cent of the total.[46] Even though these estimates are not exact, owing to a typical lack of statistics on the subject, they suffice for a comparison, with data from Japan (76 per cent), United States (69 per cent) and the twelve member countries of the EU (65 per cent).[47]

Such evidence is the basis of two other observations which epitomize what I have been saying so far: (a) there are in Argentina today only 10,000 to 15,000 researchers involved in S&T activities, depending on how these activities are measured;[48] and (b) *outside* Argentina, between 30,000 and 50,000 university-trained scientists and technicians live and work, educated in the country but forced to leave owing to the political and economic conditions which prevailed in the last decades (see data and sources in Oteiza *et al.* 1992: 55–6).

But there is more. In a comparative study in 1987, Henry Ergas (of the OECD) tried to determine why certain nations take better advantage than others of advances in knowledge. He succeeded in identifying three types of explanatory factors. One concerns *demand*, that is, the existence of a receptive and sophisticated public which presses for constant innovations. Another refers to an *industrial structure* capable of combining competition with some mechanism which enables firms to share in the financing and circulation of scientific research. The third factor is the most important, however, and encompasses innovative *inputs*: it relates to the quality of the national scientific base; to the consolidation of research institutes as true centres of excellence; and above all to the *general education of the population*. In other words, a fundamental element enabling a country to enjoy comparative advantages is the fostering of an environment of active brains and generalized productive skills through the policies of its government. This points not only to an inescapable state responsibility but, in the final analysis, to an equitable distribution of resources. This is precisely what has *not* happened in Argentina during the last two decades. The national scientific base has seriously deteriorated; there are very few research institutes which have been able to develop into real centres of excellence; and for some time now the public education system has entered a deep crisis, coupled with an increasingly regressive redistribution of income (Beccaria 1991).

Can this situation be reversed? Not unless its causes are first modified. I quote some of these and if I do not include among them the lack of organization of the S&T complex or the budgetary shortages, it is not because I consider them unimportant but because I believe they stem from the causes I do mention. First, if the activities of the S&T complex are to develop and integrate effectively into the dynamics of the SRA, the government must define a coherent and lasting development strategy.[49] A minimum requirement is devising an industrial policy, the backbone of which should be technological modernization. In other words, the PRG should furnish the necessary impetus to establish medium and long-term goals in these areas and the state should have the political will and the instruments needed to guide public and private investments in the light of such priorities.[50] The experience of INTA provides an illustration, albeit a somewhat limited one, of the benefits that can ensue from interventions of this kind.[51]

However, neither official plans nor the type of SRA that is emerging is moving in that direction – and has not done so during the last few years. On the contrary, both the government and the economically dominant sectors imagine that state contraction by means of privatization, economic liberalization and deregulation, plus monetary stability, are all that is needed to ensure the modernization of the productive apparatus. Certainly, it was essential to check mega-inflation, which had already turned into hyperinflation. Also true was that a hypertrophic state was calling for reform, that a semi-closed economic model had become unworkable, and that there was a pressing need to change regulations which were chaotic, tangled and contradictory precisely because they were the result not of design but mainly of the chronic Argentine instability. The problem lies with both the way in which these measures have been implemented and their conversion from means to ends in themselves, which entails a clear danger of throwing the baby away with the bath water.[52]

The path followed by legislation on technology transfer clearly evidences what happened. The first law on the subject was passed in 1971, and like the 'buy Argentine' law of the same period, its three objectives were: to avoid unnecessary expenditure of foreign currency, to protect local labour, and to prevent tax evasion. Under pressure from the local business community itself, represented by the General Confederation of Industry, this law was replaced in 1974 by another, which was more systematic and established better control mechanisms. Two years later the constitutional government was overthrown, the Martínez de Hoz team was installed, and in 1977 a new law was promulgated liberalizing the transfer of technology.

The swing was completed in 1981, with the approval of Law 22.426, which was based on two premises: that only the business sector was qualified to choose the technologies it needed and that the market was the best price regulator of such technologies. Following the collapse of the military dictatorship, this law (which completely liberalized transfers between unlinked parties and established very lax *ex-post* controls on transfers between linked parties) deserved strong criticism. To cite an example, payments sent abroad for technology transfers went up six times in just a few years, in spite of the ongoing industrial recession.[53] But what is most significant is that this law was not modified at all afterwards, either by the Alfonsín government or by the present one, and it still remains in force. When critics of the law were questioned about the reasons for its permanence, their answer was that in view of the profound change which has taken place in the ideological climate of the country, the risk today is that the law will simply be abolished – as already hinted by the Minister of the Economy – thus removing even the minimum guarantees it does provide. And in any case, in the absence of an industrial and technological policy, little can be done by the mere reform of a legal instrument.

A second cause relates to the 'national style of making politics', which, among other things, is essentially short-term. It is true that this has a lot to do with the frequent and serious alterations experienced by the PRG. But, unlike the situation in France, another country with a chequered political history, the problem here appears to be determined by the absence of a stable, firmly established and well-paid civil service, capable of ensuring continuity. It is worth recalling that continuity was precisely one reason for the success of the CNEA.

Unfortunately, processes of scientific and technological ripening are at odds with short-termism and hardly admit short-cuts or automatic solutions: instead they demand sustained investments and medium and long-term programmes, together with leaders who, having agreed on the goals and established all the necessary quality and suitability requirements, are prepared to wait and, even more, to tolerate the risk factor these processes entail. In general terms, this is not what has happened in Argentina, which has suffered the added burden of recurrent ideological purges that dismantled, or simply destroyed, painstakingly formed work teams. In my opinion, the short-termism, often coupled with corruption, partly explains why Argentina is one of the few countries in the world where the National Telecommunications Company (ENTEL), now privatized, neither promoted the development of its own laboratories commensurate with its size, nor used its enormous purchasing power to stimulate the growth and the consequent technological advancement of a network of local suppliers. Unlike the CNEA, it settled for the immediate security offered by two large foreign companies, which naturally made (and still make) their R&D investments in their places of origin. And it is curious that once privatization came, these large companies substantially lowered their prices when contracting with the new owners.[54]

The third factor – which I deliberately choose as a conclusion to this chapter – is the lack of a real national awareness of the decisive importance that scientific and technological activities have for the country today. Recognition of this fact has of course been a topic in the rhetoric of political leaders for some time. Yet too much specific content is lacking and, above all, the subject has not become a matter of central importance for the majority of the population. Decidedly, it is now essential to mobilize the participation and the support of the majority if relevant changes are to take place at the PRG level. In turn this should lead to a redefinition of the parameters of the SRA which is emerging. The aim must be to galvanize a collective will capable of realizing that without basic and applied research and without technological innovations, there can be no hope of a viable, sustained and equitable development. And in this process of clarification and dissemination, I believe a very great responsibility rests on the shoulders of the science and technology community itself, whose defensive withdrawal condemns it to impotence and, in quite a few cases, to extinction.

In the mid-1960s, some analysts saw the stepping-up of research in Europe as necessary for psychological, political and, in a sense, moral reasons, to prevent the Europeans from losing confidence in themselves. Applied to Argentina, today this recommendation sounds more relevant than ever.

NOTES

1 The author wishes to thank Maria Inês Bastos, Roberto Bisang, Charles Cooper, Bernardo Kosacoff, Miguel Murmis, Hugo Nochteff, Enrique Oteiza and Samuel Wangwe for their useful comments on an earlier version of this paper. For the purposes of this study, the following persons, among others, were interviewed: Carlos Maria Correa; Norberto Ferrú; Gerardo R. Gargiulo; Alejandra Herrera; Ricardo Laferriere; Enrique Oteiza and Luis A. Ravizzini. I acknowledge my gratitude to all of them, and also to Claudia D'Angeli, who assisted in conducting the interviews. I need hardly add that responsibility for the contents of this chapter is entirely my own.

2 The concept bears a close relationship to two elaborations which stem in part from similar concerns. One is that established by Aglietta (1977) which gave rise to the formation of the so-called 'French school of regulation', although in this case the term *regime of accumulation* was reserved for economic relations and *mode of regulation* for the set of institutions and patterns of behaviour that control such relations (see for example Lipietz 1987). The other elaboration comes from Gordon, Edwards and Reich (1982), who created, in turn, the category of *social structure of accumulation*. A major difference with these concepts, though, is that both give too much causal weight, in my opinion, to the form of work organization in the industrial plants. Other contributions which point in the direction I indicate in the text are those of Jessop (1983) and Block (1986, 1990). Clearly, the influence of the classic work by Polanyi (1944) hovers over all of them to a smaller or larger degree.

3 At the beginning of the century, Durkheim and Fauconnet (1903: 487) were already censuring the classical economists for having created 'a non-existent economic world, a *Guterwelt*, a world apart, that is always identical to itself and in which conflicts between purely individual forces are resolved in accordance with immutable economic laws'. They concluded in terms which are most pertinent to my argument: 'It is in fact inside collectivities, which are quite different from one another, that individuals endeavour to make money; *and both the nature and the success of these efforts change together with the nature of the collectivity in which they appear*' (my emphasis).

4 I cannot dwell on these topics other than for a very brief clarification. First, as I have already said, the difference between the moments or stages of an SRA is basically an analytical one, so that, for example, the decline of one regime could

coexist (and combine with) the emergence of another, sometimes creating the social and political tensions which are characteristic of those eras when 'the old does not finish dying and the new is not yet fully born'. It is also possible for that stage of decline to be short and/or partial, depending both on the dynamism and the form of operation of the emerging regime. It is not a question then of reifying the moments as if their succession were ineluctable, and, of course, a regime can be cut short even before reaching the stage of consolidation. Finally, the essential feature of a transitional situation is precisely that there are not enough elements present to claim that a specific SRA is in a position clearly to subordinate the whole.

5 Note that the general finding of a recent comparative study of eighteen countries where similar stabilization plans have been applied is precisely that each adjusted to them 'in different ways, depending on their own local institutions, macroeconomic structure, and relationships among major political and social groups' (Taylor 1991: v).

6 'Remove the regime of capital and the state would remain, although it might change dramatically; remove the state and the regime of capital would not last a day' (Heilbroner 1985: 105).

7 I use these terms in the same way as Heilbroner (1985: 18–19) uses them in his analysis of capitalism, in which *nature* means 'the forces or determinative agencies . . . its behavior-shaping institutions and relationships', and *logic* means the system's trajectory towards 'the pattern of configurational change generated and guided by [its nature]'.

8 It is precisely because it is necessary to consider the SRA and not just the PRG, and because their *tempos* do not coincide, that one economist who looked for possible connections between the economic decline of Argentina and the various PRGs which were successively established was obliged to conclude: 'Anyone who tries to explain the evolution of the economy on the basis of the *forms* of government presiding over each period will encounter unsurmountable difficulties' (Llach 1987: 43).

9 In this sense, its conceptual status is similar to that of the dichotomy of state and civil society. The latter is still valid in an examination of Nazism, for example, even when the frontiers between the public and private domains become very blurred; however there is no doubt it is at its clearest in the case of the nineteenth-century Mancurian liberalism.

10 So as not to complicate this brief presentation unduly, I am referring to cases where elements, which can be labelled as belonging to the SRA, seek to influence components which belong to the PRG. However, this should in no way be interpreted as meaning the pressures and influences are one-way; as is obvious, agencies identifiable as part of the PRG also try to intervene in the SRA domain, with varying degrees of success.

11 See, for example, the contrast established by Eyerman and Jamison (1990: 16) between the development of social movements in Denmark and Sweden in terms of the 'different frameworks within which political life is conceptualized'.

12 Several of the following paragraphs draw particularly on the excellent work carried out on the subject by Solberg (1987) and Adelman (1991).

13 The prairies were purchased in 1869 from the Hudson's Bay Company and became the Northwest Territories, under the direct control of the federal government. Even after Manitoba (1870), Alberta (1905) and Saskatchewan

(1905) became provinces, Ottawa maintained its rule over public lands in the prairies, a situation which continued up to 1930.

14 To appreciate the extent to which this policy entailed a conscious design and was firmly led by the state, it should be recalled that initially it came up against the liberal opposition of those who advocated north–south integration (with the United States) and were against protectionist policies; and later it antagonized the manifest free market tendencies of the grain producers, who were keen to import agricultural machinery and implements cheaper than those made locally.

15 The financial crisis of 1890 had forced the Argentine government to sell for fiscal reasons those lands in the Pampas which it still owned. Thus, at the height of the migratory process, from the middle of the 1890s to 1914, the granting of land to producers was essentially a 'public affair in Canada and a private one in Argentina' (Adelman 1991: 9). It was also during that decade that the Argentine ranchers met with success in exporting cattle on the hoof, and this required breeding improvements which would have called for major investments in artificial pastures. The solution adopted was temporary renting to farmers (particularly the immigrants), who, after working plots of about 200 hectares for a couple of years, were required to return them sown with alfalfa (Sábato 1981: 72).

16 It is particularly interesting to note that during the same period, and not by chance, the Canadian government actively sought to discourage Italian immigration into the country. 'Instead, federal authorities turned to potential migrants in Northern Europe, and especially the United States, where immigrants were both inclined to stay in the country of settlement and brought with them at least some capital to set up a farming enterprise' (Adelman 1991: 13).

17 In this case it was two minor parties, the Socialists and the Progressive Democrats, who campaigned, without much success, in favour of liberalizing the naturalization laws.

18 Not all farmers in Argentina were leaseholders of course, however much the typical production mix on the Pampas consisted of cattle ranching, leased agricultural smallholdings and temporary labour for agriculture. Nevertheless, what I am saying about the tenants is also generally true of the small owners. It is therefore fruitless to seek an explanation of the differences simply by comparing the farmers in each country. The essential element in understanding their behaviour is first to place them within the framework of the different SRAs and PRGs which developed in the two areas.

19 As usually happens in these cases, the only available alternative was extraparliamentarian protest, episodic by nature – as illustrated in 1912 by the so-called 'grito de Alcorta', an act of rebellion by the tenants which led to the creation of the Argentina Agrarian Federation.

20 For the early divorce between education and production in Argentina, see Tedesco (1982: 35–63). As this author maintains – and I am going to argue further later on – 'more than an economic function, the development of education performed an essentially political function' (ibid.: 59).

21 Quoted by Myers (1992: 91), to whose well-documented essay on the shaping of the S & T complex during the period 1850–1958 I refer the reader.

22 European influence no doubt played a part here. As the physicist Carlo Rubbia observed recently: 'In America, science is business. In Europe it is still seen as culture' (*The Economist*, 9 Jan. 1993, p. 21).

23 To get an idea of the degree to which a reorientation of the SRA was at stake, it is instructive to read the Reports for 1976 and 1977 of the Argentine Rural Society, which had already criticized the industrialist projects of Frondizi, Krieger Vasena and Peronism in the 1970s. The first of these documents hails the fact that Martínez de Hoz was seeking to replace 'the pattern of a closed and self-sufficient economic scheme by a more open one, in which efficiency and competition will play their proper roles. We share the ideas of the Minister of the Economy: Argentina will be the winner'. In 1977, the organization pursued its attack on 'an overprotected industry, walled in behind very high import tariffs', a victim of that 'entrenchment, fear of competing and insecurity of the entrepreneurs, which began as a response to the crisis of 1930 and was grounded after the Second World War on an economic philosophy which advocated self-sufficiency at any cost'.

24 'There was a revision of policies during the first three years of the 1976 military regime, but then external credit gave a shot in the arm to all the financing arrangements. During a period of sheer phantasy, everything was possible again: the financing of the fiscal deficit, availability of cheap credit, the raising of salaries and the level of activity, the spending and saving of dollars. When the bubble burst the system was bankrupted' (Canitrot 1992: 8). Graphic and partly true as it is, I believe this description underestimates the importance of that 'revision of policies' which led, for example, to the squander of foreign credit, where other countries, such as Brazil and Mexico, used it largely to expand their local production bases.

25 As Kosacoff (1987: 19) points out, between 1975 and 1982 industrial production fell by more than 20 per cent; industry's share in the GDP dropped by 21 per cent; almost a fifth of the larger manufacturing concerns closed; the share of wage earners in the sector's income fell from 49 per cent to 32.5 per cent; and since 1977, the level of investment in capital goods fell at a rate of over 5 per cent per year.

26 The huge transfer of resources from the public sector to the powerful few private concerns resulting from this had a 'profound impact in centralizing capital and concentrating the industrial markets, since the approved and completed projects [were channelled] into just a few industrial branches, and the implicit transfer of resources [benefited] just a few large companies, which are central today in the Argentine economic process' (Azpiazu and Basualdo 1988: 30). Note that just half a dozen branches involved in the production of intermediate goods (paper, cement, petrochemicals and steel) received over 50 per cent of all authorized investment.

27 In accordance with my contention in previous pages, deregulation should be understood essentially as a *change* in regulatory guidelines and not, of course, as their elimination. In this respect it is significant – although, given what I am saying, hardly surprising – that the current enthusiasm for deregulation under Minister Cavallo is driven mainly by the so-called 'law of convertibility' which absolutely *regulates* the local price of the American dollar, that is to say one of the key variables of the economy.

28 In another work I have examined some of the reasons leading to this outcome (Nun 1992). I think it is relevant here to mention at least one of these reasons. I refer to the failure of the government in identifying its adversaries. Except that in a society as conflict-ridden as that of Argentina, not to specify the identities

and the oppositions of a political project earmarked for change was not just naïve, it meant standing defenceless against dominant interest groups which certainly did take the trouble of defining the lines of cleavage. It is therefore significant that in January 1989, in its joint letter of resignation, the economic team headed by Sourrouille belatedly regretted the fact that it did not have the backing 'of a political and social majority able to mobilize sufficient power to neutralize the resistance of vested interests'; and that a few months later, the former deputy minister for the economy, Adolfo Canitrot, publicly lamented 'the failure of the Radical government in selecting its adversaries, with the result that the adversaries were the ones choosing us'.

29 For an excellent analysis of certain central features of the course to which I refer, the reader is directed to the first part of Nochteff's paper (1993).

30 See the useful regional picture provided by the introductory study to this Project, prepared by Bastos (1992). Also Oteiza *et al.* (1992: 115–25).

31 In 1984, CONICET set up the Technology Transfer Area in order to link technological research to the production sectors. A year later it established a Technology Transfer Office and, in 1986, the Advisory Commission for Technological Development, composed of researchers, business leaders and state officials.

32 Apart from CONICET, added to this is the inability of the SECYT effectively to co-ordinate the S&T complex itself. Certainly a series of institutional factors has contributed to this result, such as the fact that the various members of this complex can discuss their financing needs directly with the Secretary of the Treasury. It is in any event revealing that eight years after the SECYT was created, its present head had to admit that 'we are in need of a Law for the Sciences which will structure the national science and technology system, *which still does not exist*' (declarations by Dr Raúl Matera to *El Cronista*, Buenos Aires, 23 October 1992, emphasis added).

33 The technological change consisted first of a change in farming techniques; then, the total mechanization of agriculture, followed in the 1970s by the introduction of improved seeds and the diffusion of soya cultivation; and finally the more generalized use of agrochemicals. It can be seen that these were partly a result of modernizing operations in the productive units and partly an effect of the worldwide advance in technology which took shape after the 1960s and which, thanks to the discoveries in genetic engineering, biotechnology and chemistry, gave a boost to the production of seeds, fertilizers and pesticides (see Nun 1991 from which I quote a few passages in the text).

34 Today, a statement like this should be taken, though, with more than a grain of salt, given present tendencies to get seeds patented; to protect private rights on pesticides; etc.

35 It should be added that half the posts on the INTA's Board of Directors were reserved from the beginning for representatives of the cooperatives and producers' associations.

36 It is interesting to note that, largely owing to its own action, the context in which the INTA operates has changed considerably since the 1950s. Thus, from the middle of the last decade, a process of restructuring has begun in the organization which, among other things, emphasizes decentralization, regionalization and participation, and represents a move from a *modus operandi* which linked the technicians directly to the producers, to another in which the

aim is rather to establish increasingly stronger bonds between the institution and the local producers' organizations (Cosse 1992). Furthermore, INTA has reduced significantly its outreach activities despite criticisms from the small producers. It seems clear that such restructuring and retrenchment are due in part to the growing role played by agribusiness and their strategies.

37 To learn more about the serious threats that seem to hang over the organization's future in the framework of the Menem government's present policies, see for example the forthright statement by Enrique Martínez, ex-President of INTI (*Página 12*, 19 Nov. 1992).

38 I quote this item, and others relating to the CNEA, from the study carried out by Valeiras (1992: 130–40). The other source I use is the research by Adler (1987: 283–302), who makes a very useful comparison between the Argentine success and the Brazilian failure in seeking autonomous development in the nuclear field.

39 In 1958, the CNEA inaugurated its first research nuclear reactor, built entirely within the country and using home-produced fuel. This was unprecedented in Latin America. Later, in 1965, the CNEA assumed responsibility for carrying out its own feasibility study for what was to be the Atucha I Nuclear Power Station. Although Atucha I was built by Siemens AG through a 'turnkey' contract, an 'open package' condition was added in an appendix 'which regulated the nature of supplies and services of Argentine origin, together with a systematic examination of national industry potential and technical assistance to it to facilitate its effective participation' (Valeiras 1992: 135).

40 Summarized by Adler (1987: 291): 'Under his [Quihillalt's] guidance, the CNEA and the atomic programme developed into the most successful institution and national programme Argentina ever had: objectives were set and were partly achieved; the infrastructure for science and technology was created; human resources were developed; the road to autonomy was laid; atomic reactors and other units were built; and Atucha I was almost completed.'

41 Regrettably, this exodus is not the first one: CNEA's researchers were brutally victimized earlier, during the last military dictatorship.

42 See Sergio Emiliozzi, 'La CNEA en la mira', *Página 12*, 30 Jan. 1993. Adler himself (1987: 281), in the above-mentioned comparison between the Argentine success and the Brazilian failure in the nuclear field, sees the centralization of the CNEA and the decentralization of the CNEN as key explanations.

43 The CNEA enjoyed good financial backing from the public sector for a number of years, and it is only during the last decade that, as I have just mentioned, its finances have come under a cloud. With regard to the INTA, its legal self-sufficiency was strengthened from the beginning by the creation of the National Fund for Farming Technology, which charged 1.5 per cent on agrarian exports and maintained the organization. Today, its financing continues to be specific and automatic, but derives from the 1 per cent statistics tax charged on imports.

44 Carlos Correa rightly maintains that, given the composition of expenditures and investments, in Argentina the allocations in the public budget are a much more relevant indicator for evaluating scientific policies than technological ones since, in the latter case, the most important factors are import tariff protection; biases in the granting of loans and subsidies; the setting of industrial standards; etc. (Interview on 30 Dec. 1992). Although I refer to some of these matters in other parts of this chapter, this is a point that deserves much further elaboration

– especially now that the historical record has proved untenable the old image of a direct link between scientific developments and technical innovations.

45 See Nestor G. Gaggioli, 'Science and Development', *Revista Noticias*, 18 Oct. 1992.

46 I quote from a recent statement by the Secretary of Science and Technology, Dr Raúl Matera: 'The internal composition of the national economic effort in science and technology reveals a major distortion, *since more than 97 per cent is coming from the state sector*. This contrasts sharply with the pattern in the industrialized countries. In these, the private sector share is generally about 50 per cent' (a speech given on 15 July 1992 and reproduced in *Ciencia y Tecnología*, vol. 3, no. 28, p. 6, my emphasis). Note also that the private universities are mainly dedicated to teaching and carry out almost no research.

47 See 'Europe's technology policy' (*The Economist*, 9 Jan. 1993), in which there is also an examination of the efforts presently being made by the European Commission of the EU to almost double R&D investment.

48 The contrast with Japan, which forty years ago had a per capita income significantly lower than Argentina, is overwhelming: Japan currently has 435,000 active researchers. The EU figure is 580,000; and that of the USA, 950,000 (Data from *The Economist*, 9 Jan. 1993).

49 In the absence of such a strategy, the commendable attempts made by various institutions in the PRG to establish links with the SRA end up being quite marginal. I referred previously to efforts made in this direction by the CONICET. For its part, the University of Buenos Aires, for example, within the framework of Law 23.877 on the promotion and development of technological innovation, encouraged the creation of UBATEC with the aim of relating 'the scientific-academic system with the goods and services production sector'. UBATEC is a corporation formed by the University, the General Confederation of Industry, the Argentine Industrial Union and the Municipality of Buenos Aires.

50 In a similar vein, Kim and Dahlman (1992: 438) aptly distinguish between the direct and indirect instruments that affect scientific and technological development: 'The former shapes the direction and pace of the supply side of technology development by strengthening technological capability. It is usually referred to as science and technology policy in the narrow sense of the term. The latter shapes the demand side of technology development by creating market needs for technological change and by providing various tax and financial incentives to lubricate linkages between supply and demand. This is usually referred to as industrial policy.'

51 This is not the place to elaborate in detail on this point. It should be enough to say that I do not mean state intervention of any kind. Argentina (and the Latin American countries in general) have plenty of experience of the failure of generic industrial promotion policies which 'because they were not subject to a clear timetable for the reduction of protection and to explicit commitments by the business community on the subject of R&D, exports, etc., and in particular because there was no *ex-post* management control' ended up becoming 'a set of resolutions to establish discriminatory and badly directed subsidies, totally incapable of inducing the conduct required of the business sector' (Katz *et al.* 1986: 340).

52 Bisang (1993) has recently called attention to an important point, which further aggravates the current situation. All the main institutions of S&T that exist in Argentina were created in the wake of a different RSA, a different international context, and a different conception of the links between scientific research and technological innovation than the one prevailing today. All this generates institutional resistances to adapt to the new environment in the making and tends to misdirect, at least in part, the ever-more scarce resources available for S&T.

53 See, for example, the interesting contribution by Carlos M.Correa and Luis A. Ravizzini to the debate on this topic promoted in 1986 by the magazine *Realidad Económica*, no. 73, pp. 92–127.

54 For documented analyses of the painful trajectory of the telecommunications sector in Argentina, see Herrera (1987, 1989), Herrera and Petrazzini (1992) and Di Benedetto and Herrera (1992). For a detailed account of the disastrous privatization process, see Verbitsky (1991: particularly 197: 201–9; 212–22, 235–8 and 253–65).

REFERENCES

Adelman, J. (1991) 'The social basis of technical change: mechanization and the wheatlands of Argentina and Canada, 1890–1914, mimeo.

Adler, E. (1987) *The Power of Ideology. The Quest for Technological Autonomy in Argentina and Brazil*, Berkeley, University of California Press.

Aglietta, M. (1977) *Regulation et crise du capitalisme*, Paris, Calmon-Levy.

Arnaudo, A. A. (1987) *Cincuenta años de politica financiera argentina (1934–1983)*, Buenos Aires, El Ateneo.

Azpiazu, D. and Basualdo, E. M. (1988) *Cara y contracara de los grupos economicos. Crisis del estado y promoción industrial*, Buenos Aires, Cantaro.

Azpiazu, D. (1992) 'Asignación de recursos públicos en el Complejo Científico y Tecnológico. Analisis del presupuesto nacional', in Enrique Oteiza *et al. La política de investigación científica y tecnológica argentina. Historia y perspectivas*. Beunos Aires, Centro Editor de America Latina.

Bachrach, P. and Baratz, M. S. (1970) *Power and Poverty*, New York, Oxford University Press.

Barsky, O. (1988) 'La caída de la producción agricola pampeana en la década de 1940', in Osvaldo Barsky *et al.*, *La agricultura pampeana: transformaciones productivas y sociales*, Buenos Aires, Fondo de Cultura Economica.

Bastos, M. I. (1992) 'The politics of science and technology policy in Latin America', Maastricht, UNU/INTECH (mimeo).

Beccaria, L. A. (1991) 'Distribución del ingreso en la Argentina: explorando lo sucedido desde mediados de los setenta', *Desarrollo economico*, vol. 31, no. 123, pp. 319–38.

Bisang, R. (1993) 'Industrialización e incorporacion del progreso técnico', Buenos Aires (mimeo).

Block, F. (1986) 'Political choice and the multiple "Logics" of capital', *Theory and Society*, no. 15, pp. 175–92.

Block, F. (1990) *Postindustrial Possibilities*, Berkeley: University of California Press.

Caldelari, M., Casalet, M. Fernández, E. and Oteiza, E. (1992) 'Instituciones de promoción y gobierno de las actividades de investigación', in E. Oteiza *et al.* *La política de investigación científica y tecnológica argentina. Historia y perspectivas*, Buenos Aires, Centro Editor de America Latina.

Canitrot, A. (1992) 'La destrucción del estado argentino y los intentos posteriores de reconstrucción', Buenos Aires, Fundacion Simon Rodrigues/CEDES.

CGE (1990) *Estratégia para el crecimiento con equidad*, Buenos Aires, Instituto de Investigaciones Economicas.

Cosse, G. (1992) 'El aparato de extension del INTA', in Osvalso Barsky *et al.*, *El desarrollo agropecuário pampeano*, Buenos Aires, Fondo de Cultura Economica.

Di Benedetto, L. and Herrera, A. (1992) 'Telecomunicaciones', in Enrique Oteiza *et al.* *La política de investigación científica . . .*, Buenos Aires, Centro Editor de America Latina.

Durkheim, E. and Fauconnet, P. (1903) 'Sociologie et sciences sociales', *Revue Philosophique*, no. 55, pp. 465–97.

Evans, P. (1990) 'The State as a problem and solution: predation, embedded autonomy and structural change', in Stephan Haggard and Robert Kaufman (eds), *The Politics of Adjustment. International Constraints, Distributive Conflicts, and the State*, Princeton, NJ, Princeton University Press.

Eyerman, R. and Jamison, A. (1990) 'Social movements: contemporary debates', *Research Reports*, Dept. of Sociology, Lund University.

Gordon, D., Edwards, R. and Reich, J. (1982) *Segmented Work, Divided Workers*, New York, Cambridge University Press.

Gourevitch, P. (1986) *Politics in Hard Times*, Ithaca, NY, Cornell University Press.

Heilbroner, R. L. (1985) *The Nature and Logic of Capitalism*, New York, W. W. Norton.

Herrera, A. (1987) 'Telecomunicaciones: reestructuración productiva y empleo en la Republica Argentina', *Desarrollo Económico*, no. 27, pp. 105, 107–27.

Herrera, A. (1989) *La revolución tecnológica y la telefonia argentina*, Buenos Aires, Legasa.

Herrera, A. and Petrazzini, B. A. (1992) 'Revolución tecnológica, re-regulación y privatización: alcances y limites del milagro. El caso argentino', Buenos Aires, mimeo.

Jessop, B. (1983) 'Accumulation strategies and hegemonic projects', *Kapitalistate*, vol. 10/11, pp. 89–112.

Katz, J. M. (n.d.) 'Reflexiones acerca de la relación entre la capacidad tecnológica interna, acumulación y productividad industrial', Buenos Aires, mimeo.

Katz, J. M. *et al.* (1986) *Desarrollo y crisis de la capacidad tecnológica latinoamericana. El caso de la indústria metalmecanica*, Buenos Aires, IDES.

Kim, L. and Dahlman, C. J. (1992) 'Technology policy for industrialization: an integrative framework and Korea's experience', *Research Policy*, no. 21, pp. 437–52.

Kosacoff, B. (1987) *Desarrollo industrial e inestabilidad macroeconomica. La experiencia argentina reciente*, Buenos Aires, CEPAL.

Lattuada, M. (1988) *Política agraria y partidos políticos, 1946–1983*, Buenos Aires, Centro Editor de America Latina.

Lindblom, C. E. (1977) *Politics and Markets*, New York, Basic Books.

Lipietz, A. (1987) 'Rebel sons: the regulation school', *French Politics and Society*, no. 4, pp. 17–26, Cambridge, Mass.

Llach, J. J. (1987) *Reconstrucción y estancamiento*, Buenos Aires, Tesis.

Myers, J. (1992) 'Antecedentes de la conformación del Complejo Científico y Tecnológico, 1850–1958', in Enrique Oteiza *et al.*

Nochteff, H. (1991) 'Reestructuracion industrial en la Argentina: regresión estructural e insuficiencia de los enfoques predominantes', *Desarrollo Economico*, no. 31, pp. 338–58.

Nochteff, H. (1993) 'Informe para el Seminario del INTECH en Maastricht', Buenos Aires, mimeo.

Nun, J. (1987) 'La teoria política y la transición democrática', in José Nun and J. C. Portantiero, *Ensayos sobre la transición democrática en la Argentina*, Buenos Aires, Puntosur.

Nun, J. (1989) *La rebelion del coro*, Buenos Aires, Nueva Vision.

Nun, J. (1991) 'Las promesas reformistas', in José Nun and and M. Lattuada *El gobierno de Alfonsín y las corporaciones agrarias*, Buenos Aires, Manantial.

Nun, J. and Lattuada, M. (1991) *El gobierno de Alfonsín y las corporaciones agrarias*, Buenos Aires, Manantial.

Obschatko, E. S. (1988) 'Las etapas del cambio tecnológico', in O. Barsky *et al. La agricultura pampeana: transformaciones productivas y sociales*, Buenos Aires, Fondo de Cultura Economica.

O'Connell, A. (1985) 'La economía argentina: situación y perspectiva', Buenos Aires, mimeo.

Oteiza, E, *et al.* (1992) *La política de investigación científica y tecnológica argentina. Historia y perspectivas*, Buenos Aires, Centro Editor de America Latina.

Piñeiro, M. and Trigo, E. (1983) 'Towards an Interpretation of Technological Change in Latin American Agriculture', in Martin Piñeiro and Eduardo Trigo (eds) *Technical Change and Social Conflict in Agriculture*, Boulder, Col., Westview Press.

Polanyi, K. (1944) *The Great Transformation*, New York, Beacon Press.

Sábato, J. F. (1981) *La pampa pródiga: claves de una frustración*, Buenos Aires, CISEA.

Solberg, C. E. (1987) *The Prairies and the Pampas: Agrarian Policy in Canada and Argentina*, 1880–1930, Stanford, Stanford University Press.

Taylor, L. (1991) *Varieties of Stabilization Experience*, Oxford, Oxford University Press.

Torre, J. C. (1983) *Los sindicatos en el gobierno, 1973–1976*, Buenos Aires, CEAL.

Tedesco, J. C. (1982) *Educación y sociedade en la Argentina (1880–1900)*, Buenos Aires, Centro Editor de America Latina.

Valeiras, J. (1992) 'Principales instituciones especializadas en investigación y extensión', in Enrique Oteiza *et al. La política de investigación científica . . .*, Buenos Aires, Fondo de Cultura Economica.

Verbitsky, H. (1991) *Robo para la corona*, Buenos Aires, Planeta.

Vogel, D. (1986) *National Styles of Regulation. Environmental Policy in Great Britain and the United States*, Ithaca, Cornell University Press.

Zysman, J. (1983) *Governments, Markets, and Growth*, Ithaca, Cornell University Press.

3 State autonomy and capacity for S&T policy design and implementation in Brazil

Maria Inês Bastos[1]

Brazil has been experimenting with S&T policy since the late 1960s when the government first defined the incorporation of S&T into the productive system as a strategic goal to be attained through real technology transfer and the development of the Brazilian capacity to innovate. This quarter of a century-long experimentation has resulted in the creation of a complex set of institutions and organizations which regulate the process of technology transfer, provide skilled human resources, and fund and perform research activities. While relatively efficient in supplying skilled labour and high-level human resources for research, the policy has managed to stimulate innovation capabilities at firm level in only the few cases in which it is integrated with sectoral industrial policies. Beyond the sectoral level, technology policy has struggled to provide the general conditions for technology development. The regulation of technology transfer and the granting of protection to industrial and intellectual property rights has been set up and adapted to changing domestic and external conditions. Incentives and subsidies have been established to stimulate technological development at firm level but in general they do not seem to have been effective. Human resources for research and high-level technical personnel have been provided, but not in the amount, quality and specializations required. In addition, a high illiteracy rate and an inefficient system of primary education have constituted general constraints to innovation.

Brazilian public policies for industrialization, while creating a diversified industrial sector have, however, considerably limited domestic capability for industrial innovation. Industrialization has been heavily based on foreign capital and technology while industrial production has been strongly subsidized with the supply of Brazil's large market absorbing the majority of such production. The latter factor has orientated the conception of product and process technology. Industrialization has been part of a development model based on a regressive profile of income distribution, a demand by higher-income groups for luxury products, and an increasing foreign debt. The model has contributed to social marginality and the poor utilization of manpower resources. Within this development model, S&T policy has been powerless to overcome reduced market stimuli for innovation. The exhaustion of the import-substitution strategy in Brazil and its inability to provide the conditions for sustained development became widely recognized in the 1980s when the external indebtedness and major resource demands for the servicing of the foreign debt proved the limited capacity of Brazil's industry to compete via innovation and product diversification. S&T policy in Brazil has been unable to overcome the constraints that deter industrial innovation at firm level and diffusion of new techniques within the economy as a whole. S&T institutions and organizations in Brazil lack the participation of private firms and systemic linkages, and are of diverse quality. Furthermore, they have been weakened by economic crisis.[2]

A comprehensive explanation of the relatively poor results of state intervention to incorporate S&T in the productive system through real technology transfer and the development of capacity to innovate in Brazil is far beyond the limits of this chapter. Instead, I want to explore only some of the political dimensions of this complex phenomenon. From a political sociological perspective, I will discuss S&T policy in Brazil centred on state capacity for policy design, implementation and change.[3] My central argument in this chapter is that the Brazilian state has shown a better capacity for S&T policy design than for its implementation, due to the characteristics of state autonomy in Brazil. If policy design is dependent on state 'expertise', which can be used by an autonomous state even when links with societal forces are weak or unstable, policy implementation depends not so much on the capacity to articulate administrative means but on the capacity to link with societal forces. In this respect, the autonomy of the Brazilian authoritarian state was more a liability than an asset for S&T policy-making. In the absence of strong institutions, the autonomy of a legitimate government does not guarantee efficiency.

The chapter is divided into three sections. Section 1 contains a discussion of domestic and foreign political determinants of developing states' capacity for policy design, implementation and change. Section 2 presents the analysis of state capacity for S&T policy design and implementation in Brazil. In sub-section 2.1 the concept of S&T policy is presented. Sub-section 2.2 contains a brief discussion of societal forces and Brazilian import-substitution policies. Sub-section 2.3 presents an analysis of societal forces and S&T policy in Brazil. Finally, in sub-section 2.4 the liberalization and modernization project introduced in 1990 is discussed. Section 3 presents the conclusions of the chapter.

1 FOREIGN AND DOMESTIC POLITICAL DETERMINANTS OF DEVELOPING STATES' CAPACITY FOR POLICY DESIGN, IMPLEMENTATION AND CHANGE

S&T policy design and change, particularly in developing countries, cannot be fully understood without considering the role of international economic and political incentives and/or constraints. A variety of factors – international market pressures in the form of price shocks; restriction or facility of access to markets, supplies and capital; political conflicts with trade partners which can alter patterns of trade and investment and give rise to external shocks – are all relevant elements affecting S&T policy design and change.

Political pressure from external political actors or by the international system has been analysed as a significant explanation of change in industrialization policies (Haggard 1990) as well as adjustment policies (Nelson 1990), and of high-tech policies (Bastos 1991, 1992, 1994). Haggard (1990) showed that major powers have influenced industrialization policies of developing countries either in the extreme case of formal empires, or through military alliance and economic hegemony. The present author explored international pressures for policy change linked to economic hegemony and expressed in trade conflicts, showing how trade sanctions affect the balance of domestic support to policy (1991, 1992, 1994). Analysing how extensive conditionality and co-ordination among major financing agencies produce external pressure on internal policies, Nelson (1990) pointed out that these pressures have been unprecedented in scope, detail and number of countries affected. The resulting adjustment policies have direct consequences on state capacity for creative S&T intervention. These macroeconomic policies affect not only the amount of public and private resources available for investment in S&T, but also the political underlying coalition and the organizational structure for the implementation of S&T policy. In addition, government commitment and social interest in scientific and technological development are deeply affected by the adjustment effort by which short-term stabilization concerns tend to dominate over long-term goals (Cassiolato, Hewitt and Schmitz 1992). Additional constraints on state capacity for design and implementation of technology policies, particularly the current new technologies, come from the frantic pace of international technological innovation in these new areas, which turns technological goals into moving targets.

At domestic level, state capacity for policy-making is decisively dependent on the nature of its relationship with societal forces. The relationship between state and society can be apprehended on two theoretical levels. One refers to the pact of domination involving the most powerful classes or class fractions at the basis of the capitalist state. Changes in this domination pact help explain the political conditions for the implementation of industrialization policies (Kaufman 1990; Haggard 1990). At this theoretical level, changes in the relationship between state and society are seen over larger periods and explain general historical trends. Import-substituting industrialization strategy has been explained by shifts in the composition of politically and economically dominant groups and their relationship with governments (Fishlow 1987; Kaufman 1990; Haggard 1990). Focus on the pact of domination has also drawn attention to internal conflicts of interests and their expression in policy consistency. At this level, the analysis paints a background picture of the general strategic priorities for state intervention according to which S&T policies could be defined without fundamental challenge. The pact of domination determines the boundaries within which S&T policies could be legitimately defined, that is, it sets a basic limit for state autonomy. Therefore, autonomy of the state involves

> not so much the coalitional origins of general strategic priorities as the capacity to translate these general priorities into a coherent, operational programme for development which will pass without challenge within the limits defined by the existing coalitional 'pact of domination'. Such bounded autonomy provides the basis essential for developmentalist intervention and economic reconstruction where otherwise class or organizational forces might seek to abort strategic restructuring.
>
> (Deyo 1987: 230)

On another theoretical level, the relationship between state and society can also be apprehended in relation to particular policies. In the case of S&T policies this would involve the analysis of how the state takes into account the interests of producers and users of knowledge and technology. This relationship can be explored in terms of public/private co-operation, as a process of consensus building (Pack and Westphal 1986; Wade 1990) or in terms of policy 'embeddedness' (Evans 1989, 1992; Onis 1991) and is particularly relevant to the analysis of state capacity of S&T policy implementation and change.

Although this co-operation is necessary, it needs to be balanced so that the state can ensure its autonomy and prevent its capitulation to private interests. By bureaucratic autonomy is meant the capacity of independent formulation of its own goals and the ability to count on those who work within the state to perceive implementing these goals as important to their individual careers (Evans 1992: 154). Discussing the Japanese model of a developmental state, Evans (1992) considers this bureaucratic autonomy necessary for policy design, but stresses that because policy design demands accurate intelligence, inventiveness and attention to changing economic reality, it is dependent not only on expertise but also on strong links with society. On the other hand, more necessary than bureaucratic autonomy is the assurance that once goals are selected, they are accomplished. Policy implementation depends on close private/public co-operation. However, in order to be able to change policy, i.e. to 'phase out' some government support, penetration by societal forces has to be kept within certain limits so that bureaucratic autonomy can be preserved. Thus, state capacity of policy design, implementation and change depends on its 'embedded autonomy' (Evans 1992).

The concept of 'embeddedness' refers to the social relations constraints that are pervasive in all behaviour and institutions (Granovetter 1985). State institutions cannot be completely independent of social relations; for, even when autonomy isolates the state from society, 'embeddedness' is reduced to a minimum but not eliminated. 'Embedded autonomy' of the state is a concept that draws attention to the fact that enhanced relationship with social groups is not only a way for the state to fulfil its guardian role concerning the interests of the whole society, but also a way to get privately oriented social groups to collaborate towards the attainment of public goals.

Embedded autonomy depends on the development of two components. One refers to the state bureaucracy and the other to the mechanisms of articulation with society. A cohesive state bureaucracy able to define its own socially relevant goals and create the instruments to attain them is a condition required for autonomy and efficiency (Johnson 1981; Rueschemeyer and Evans 1985; Wade 1990; Evans 1989, 1992). Cohesion and coherence of state bureaucracy is highly affected by its internal dynamics expressed by recruitment patterns, career paths and corporate culture. Autonomy increases when state bureaucrats are recruited through universal criteria which assess their ability to perform corresponding duties; when satisfactory material reward makes bureaucrats less vulnerable to the exchange of public means for private gains; and when a well-defined career path makes mobility less dependent on political influence and personal connections. Recruitment patterns and bureaucratic career paths affect internal coherence not only because of their implications for the strengthening of bureaucratic expertise but also because of their contribution to the constitution of corporate culture. An internally cohesive and coherent state bureaucracy builds a public image that facilitates autonomous embeddedness.

Administrative and organizational characteristics of the decision-making apparatus, therefore, affect state capacity of policy design and implementation. Cohesiveness of the decision-making structure is a relevant point to take into account in the analysis of S&T policy where the state is internally divided. In such a situation, disarticulation of policies is most likely to happen because in an internally divided state, each 'piece' pursues its own specific targets with a relatively large degree of independence. Power is unevenly distributed among the component 'pieces' (Abranches 1978). There is usually an inner circle of power where the top political elite in charge of making strategic decisions is concentrated. The easier the access to this inner circle by state managers, the more effective their power. Managers of S&T policy are not likely to have easy access to this circle, especially when compared with their colleagues in charge of the economic policy. S&T state bureaucracy may be kept away from the inner circles of power and isolated from the other 'pieces' of the divided state. Administrative reforms or the creation of ministries of science and technology may not change the amount of effective power at the disposal of S&T policy managers. As a consequence, the relatively smaller effective power attached to the S&T policy-making apparatus affects their ability to: (a) induce the commitment of significant parts of the administration to S&T goals; and (b) ensure the allocation of appropriate resources for graduate education, vocational and technical training, infrastructure and maintenance of research centres, financing research projects, and the provision of incentives to firm-based R&D activities. In an internally divided state, where S&T policy is treated as a 'sector' and administratively confined to a few agencies, conditions for consistent intervention are very reduced because 'state actors tend to view solutions to particular problems through the lens of the instruments that are available to them; their options are limited or expanded by the tools they have at hand' (Haggard 1990: 46).

The other component of embedded autonomy refers to ways of articulation of societal interests. A balance has to be struck between the two extreme cases – complete isolation and capitulation to societal interests. A way to facilitate this balance and enhance embedded autonomy is to build institutionalized mechanisms representing interests and recognizing social accountability of the state bureaucracy. Stable and institutionalized mechanisms of interest representation where participation of interested parties is defined according to universal criteria is superior in their effects upon embeddedness to ad hoc and personalized links. The superiority of these mechanisms is due to their effects on generating mutual binding commitment. Outcomes of ad hoc, personalized relationships may be seen as arbitrary and social groups or state bureaucrats may regard them with mistrust. Programmatic political parties, a representative electoral process, and an active and powerful legislature, are significant mechanisms of articulation of interests in a representative democracy. Legislature power and a strong judicature are prerequisites for social accountability of state bureaucracy.

2 STATE-SOCIETY RELATIONS, DEVELOPMENT MODEL, AND S&T POLICY DESIGN, IMPLEMENTATION AND CHANGE IN BRAZIL

After presenting the concept of S&T policy used in this chapter, the process of policy design, implementation and change in Brazil is discussed. Three questions orient this discussion:

1. What particular combination of interests within Brazilian society led to the definition of industrialization policies and how much space did these interests create for the emergence of S&T policy?
2. Why, in this context, was a policy targeting autonomous technological development formulated?
3. What has changed in the Brazilian society to explain current liberalization and modernization policies in substitution to the previous development model?

2.1 A concept of S&T policy

The expression 'S&T policy', very popular in Latin America and widely used in Brazil, elicits however a fundamental misunderstanding with many practical implications. Considering that science and technology, although closely related, refer to different categories of knowledge resulting from substantially different development processes, one might expect the expression to denote policies that focus only on the area of intersection of those areas of knowledge. This is not, however, the case. It has been applied to the two policy areas as if they could be treated in the same way and by the same policy instruments and within the same governmental agencies.

This situation, which is not peculiar to Brazil or even the Latin American region,[4] reflects not only a linear conception of the relationship between science and technology in the innovation process, but also the historical evolution of state intervention in the area. Technology policy was conceived as a sub-set of science policy with the underlying assumption that 'basic research creates knowledge that subsequently is incorporated into technological practice and commercial products and processes' (Mowery 1994: 9). Even after such a linear model of innovation was proved misleading (Kline and Rosenberg 1986), the idea that technological advance depends on the progress of science still remained a not quite clear assumption in the S&T policy in Brazil. This fact can also be explained by the institutional evolution of governmental support to the two policy areas. A technology policy of some kind implied by import-substituting industrialization policies was implemented by those agencies in charge of the Brazilian economic policy, while support for basic research and the training of scientists and engineers was a separate activity under the strong influence of the academic community and under the management of agencies much more linked to education. Since the mid-1970s, however, the two policy areas became institutionally fused, in formal terms, under the co-ordination of the agency that had a long experience in supporting basic research.

While science policy's aim is to develop basic research capabilities, technology policies are those 'intended to influence the decisions of firms to develop, commercialize, and adopt new technologies' (Mowery 1994: 8). The awareness that technology policy emphasizes the technological performance of firms has been particularly noticed in the recent approach to such policy in Brazil (Bell and Cassiolato 1993). The intentionality of technology policy is a crucial component of this concept because many other policies that affect firms' decisions on innovation and adoption are not necessarily designed with the purpose of influencing innovative performance. Policies that have effects on innovative performance but lack this element of intentionality can be considered as 'indirect' or 'implicit' technology policies.

Technology policy is, therefore, part of the set of public policies for innovation as it targets the development of a capability 'to generate and commercialize new and better products and production processes' (Dosi, Pavitt and Soete 1990: 3). It contributes to building 'national innovation systems', that is, the network of private and public organizations that fund and perform R&D, translate the results of such activities into commercial innovations, and affect the diffusion of new technologies within the economy (Freeman 1988; Lundvall 1988; Dahlman and Frischtak 1990; Mowery 1992).

So conceived, the definition of the boundaries of technology policy is not a clear-cut matter. Particularly difficult, though necessary, is to differentiate it from economic and industrial policies, on the one hand, and from science and education policies, on the other. 'Adoption-oriented' or 'demand' technology policies include measures and instruments belonging to economic, trade and industrial policies such as financial subsidies, provision of information via extension and demonstrations, conditions for foreign technology transfer, technical standards, regulation of intellectual property protection and government procurement (Mowery 1994; Kim and Dahlman 1992). Blurred boundaries between technology policy and science and education policies occur in relation to 'supply' technology policies by which the construction of research infrastructure and the development of basic technological capabilities are targeted. This happens not only because some of the R&D activities are performed in research institutes devoted to basic research and it is in these laboratories that the training of scientists and engineers – one important condition for the diffusion of innovation – is made. No absorption of new technology is really made possible without the necessary training of the workforce and even of the customers at large, which involves a concern with vocational training and the improvement of the general level of education. In this respect, governmental policies aim at developing domestic technological capabilities, an important element of dynamic comparative advantage.

With these important caveats, the expression 'S&T policy' is here used to bring together this complex set of technological, basic research and training activities that are part of the governmental intention of attaining the two interrelated goals of influencing the development of domestic technological capabilities and firms' innovative performance.

2.2 Societal forces and Brazilian import-substituting industrialization policy

Industrialization in Brazil was not primarily derived from state initiative, nor was import-substituting strategy in its origin a choice made by Brazilian policy-makers. The fundamental push originated from the foreign trade crisis and the subsequent 'strategy' is more appropriately seen as a result of adjustment policies to the Great Depression and lack of foreign financing (Corbo 1992). Political change in Brazil was also related to that economic crisis. The stabilizing force of the Brazilian state, the coffee oligarchy, was dislocated by the crisis, altering the balance of regional political forces that had sustained political stability. The power gap filled in by Vargas and supported by an alliance of traditional political elite and the military, implied little change in the power structure (Furtado 1972). The 1930 revolution against the Old Republic had placed civilian officials and military elite in a position to arbitrate the conflicts between the Paulista coffee planters, their regional landowning rivals, and middle-class groups in the more modernized southern states (Kaufman 1990). This mediating role gave the federal officials and military officers an increasing measure of independence from any of the competing regional oligarchies and this allowed the state to extract resources from the exporting sector and allocate them to new industrial and infrastructural activities. The continuing power of the rural elite, not its weakness, determined the choice of indirect measures such as the exchange rate and commercial policy instruments to tax the rural sector and simultaneously redistribute the proceeds to the new industries (Fishlow 1987).

Import-substitution industrialization, therefore, was initiated in Brazil by a highly conservative constellation of oligarchical and governmental forces. During the 1930s, industrial production had high growth rates. While overvalued exchange favoured the importation of capital goods for the new domestic industries, government expenditure on coffee stockpiles continued to benefit commodity exporters. Industrialism had not yet become an ideology. On the contrary governmental officials, including Vargas himself, continued throughout the decade to warn against the risks of changing Brazil's 'natural' role as a producer of raw materials (Kaufman 1990: 119). On the other hand, all upper-class groups gained from the government's efforts to demobilize and control recurrent labour militancy. This was sought through a considerable application of force in the mid-1930s, and subsequently through the institution of the corporatist framework in 1937.

After the Second World War, Brazil found itself with substantial foreign reserves in the central bank and there were positive indications of an expansion in world trade. The stage was set for a reduction in the high-level of protection for import-competing industries and in the discrimination against exportables. Continued emphasis on import substitution after the Second World War, rather than on export-promoting industrialization, was due to two political factors. One was an understandable scepticism concerning the stability of international trade as a stimulus for growth. The other was that export promotion in a resource-rich country such as Brazil would necessarily mean an emphasis upon the primary sector, with a consequent strengthening of the traditional rural elite whose influence industrialization was supposed to diminish (Fishlow 1987). The Brazilian developmental state remained inward-looking not only as an expression of its autonomous commitment to industrialization, but also as a result of the rise of an urban society organized around the industrial and public sectors. Protectionism meant support of industrial entrepreneurs and workers and a white-collar service sector (Fishlow 1987; Corbo 1992). The persistence of import substitution is also related in part to the ability to pay for it, that is, the ample availability of traditional natural resources (Kaufman 1990). The compounded effect of social and natural conditions was continued protection for existing manufacturing companies and an extension of the import-substituting process into consumer durables. Manufacturing output achieved substantial growth, but started to decline by the late 1950s and early 1960s when the easy import-substituting phase was completed.

The deepening of import-substituting industrialization in the 1960s and 1970s may be understood in connection with a change in the balance of societal forces in Brazil, in which the interests of foreign industrial capital were central, as well as the growing influence of industrialism and nationalism as dominant beliefs within state bureaucracy. A shift away from protectionism was discouraged by the decline of interests of commodity exporters in favour of foreign and domestic industrial capitalists willing to preserve or gain access to protected markets (Kaufman 1990). The influence of the ECLA doctrine among nationalist state bureaucrats played a significant role in justifying this continuity. The analysis of this role is, however, beyond the limits of this chapter. Here it is enough to point out that the lasting influence of the ECLA doctrine was mostly noted in the premises of regional political debate where nationalism was equated with protectionism (Fishlow 1987). Nevertheless, the consumer durable industry developed under its shelter was heavily based on foreign investment.

The 'triple alliance' that sustained military rule in Brazil did not completely eliminate the influence of landed oligarchical interests represented by traditional commodity exporters. These forces had retained some, albeit declining, power in the coalition which sustained industrialization so far. Their interests were in some measure deeply hurt by the military movement which championed the end of state patrimonialism that had nurtured the political power of these oligarchical groups. However, they were never really defeated and kept some position in the dominant coalition. It is necessary not to forget their participation in the movement that overthrew Goulart and the leading role they played in the political mobilization that had prepared the coup. This participation was rewarded with military retreat from a proposed land reform, and during the whole period of authoritarian rule (also extended to Sarney's transition government) they managed to block every attempt at changing the land structure in rural Brazil.

Summing up, the launching of import-substituting industrialization policies and its subsequent deepening was made politically possible because of a particular composition of the pact of domination. In the 1930s, the government gained a degree of autonomy over the landed oligarchies. After the Second World War, the composition of the dominant coalition changed: commodity exporters were still powerful, but the industrialists and their surrounding urban groups became socially and politically stronger. While domestic industrialists had grown under the protection afforded to local manufacture of substitutes of imported light consumer goods, in the 1950s policies for substitution of imported consumer durables created incentives for the foreign entrepreneurs best equipped with the required capital and technology. The interests of foreign subsidiaries became compatible with internal prosperity; this was analysed as the basis of the Brazilian model of associated-dependent development (Cardoso 1977). One important feature of this model was the development of strong protectionist business interests.

From the early 1930s to the mid-1960s, the priorities that shaped industrial policy supported by the political coalition of industrialists, commodity exporters and functional middle-class groups, did not create the conditions for the emergence of S&T policy. It is also necessary to consider that the technological requirements of the first phases of industrialization in Brazil were limited to capital goods and human resources requiring certain skills. Importation of capital goods, utilization of skilled migrants, and vocational and technical training supplied the technological demand in these first phases. Governmental intervention which resulted in the provision of these requirements did not take the form of a policy intentionally directed to reach any determined target of technological development, but was responsive to systemic demands (Guimarães and Ford 1975). In spite of that, the foundation of research institutions and the creation of governmental agencies and financing instruments for the promotion of technology acquisition in the period established the basis for future policy.[5] The priorities originating from the coalition that sustained the military coup admitted a translation into S&T policy. An expansion of industrial production to include more sophisticated products brought about a corresponding evolution of technological requirements. The mere purchase of capital goods was not a sufficient source of supply of necessary technology since industrial companies, at the time, needed more than simple operating instructions given by capital goods manufacturers. Conditions for technology transfer agreements were therefore regulated so that blueprints, consultancy and technical assistance could provide the means for specific problem-solving activities. Licensing of proprietary technology had to be negotiated. Moreover, the deepening of import substitution required domestic technological capability to select, absorb and adapt imported technology. Substitution of capital goods extended these technological requirements to the development of domestic capability of technological innovation. The objective conditions for the emergence of S&T policy were present. The translation of them into a governmental commitment was made possible because of their linkage with the fundamental interests of the dominant political coalition. However, as I will show in the following sub-section, governmental commitment to S&T policy was not without internal conflicts expressed by the policy's insulation within the development policy. Moreover, its establishment by an authoritarian state had major implications for policy implementation.

2.3 S&T policy as an act of an autonomous state

Up to the late 1960s, S&T policy had not been a subject for intentional and systematic intervention by the Brazilian state. In 1968, the Strategic Development Programme (1968/70) defined the incorporation of S&T into the productive system as a strategic goal. The programme established the need to diversify the sources of industrial growth by deepening import substitution, expanding the domestic market, and increasing export promotion. In order to achieve this diversification, particularly in relation to export promotion, the programme stated the requirement of improved competitiveness of Brazilian industry, starting with an increase in their levels of efficiency, the expansion of some dynamic industries, and reorganization and modernization of traditional industries (Guimaraes and Ford 1975: 409). In addition to the modernization of industrial production and the improvement of international competitiveness of Brazilian industries, the commitment to S&T goals as stated in the programme included the development of domestic technological capability.

> The substitution of imported industrial goods, through the intensity of the post-war period, is not sufficient to ensure a sustainable development, particularly due to its implications for market expansion and appropriateness of technology in use. It will be necessary to complement it [the substitution of imported industrial goods] with *substitution of technology*, meaning adaptation of imported technology and gradual establishment of an autonomous process of technological progress. *One can hardly find any country that had experienced a rapid and self-sustained growth without the support of an internal process of technological development.*
>
> (Ministério do Planejamento e Coordenação Geral, *Programa Estratégico de Desenvolvimento, 1968–70*, vol. I, Part II: IV–8, original emphasis, my translation)

S&T policy emerged with an outlook towards an autonomous technological development, that is, the improvement of domestic capability of complementing imported technology through adaptation and innovation.

Why did the Brazilian state take the initiative of designing such S&T policy in the first place? Economic and political arguments had been advanced to explain the initiation of the Brazilian S&T policy.[6] The economic argument stressed the need to relieve the shortage of high-skilled labour in a new phase of import-substituting industrialization and the need to cut back payments for imported technology and capital goods in order to reduce deficits in the balance of payments. The political argument emphasized ideological components: the military conception of Brazil as an emerging power, also the centrality of 'technical' knowledge as a basis for legitimating power during military rule in Brazil. In addition, it advanced the hypothesis that the initiative of launching an S&T policy in Brazil had its roots in the internal division of the Brazilian state where a nationalist segment of the bureaucracy, in alliance with nationalist military and segments of the S&T community, had enough autonomy to establish it. This convergence of interests found favourable conditions in the high economic growth rates of the period. Therefore, S&T policy in Brazil started as an independent state act, while domestic industrialists remained indifferent for a long time until they started to reap the benefits of the financial incentives created by the policy (Erber 1979). Moreover, there were various international programmes of technical co-operation for the diffusion of planning techniques including governmental support of S&T (Amadeo 1978; Cassiolato *et al.* 1981).

Each of the above explanations contributes to the understanding of the political dimensions of the initiation of S&T policy in Brazil by an authoritarian regime. None of them can be taken in isolation as a satisfactory explanation and all of them taken together still miss a few interesting aspects of the phenomenon. I want to explore here implications of the fact that such policies were autonomously defined by the state.

S&T policy in Brazil is now twenty-five years old. More than half of this experience, and certainly the most creative of it, was made under military rule and in a juncture of high economic growth rates. An overview of Brazil's S&T policy will show that since the last military government and the reduction of economic growth to stagnation, the policy has lost momentum and governmental support for S&T activities has declined to the current very low level. Starting with the early years of the 1980 decade, S&T policy institutions have become 'empty shells'. The decisive initial years of its establishment and institutional development occurred in the darkest period of political repression in the country. The policy may be said to have been a product of a government that enjoyed almost absolute power and a high degree of isolation from societal interests. What implications do these facts have? Insulation has been assumed to be a condition for autonomy and effective state intervention. In the Brazilian case, while S&T policies were designed and implemented in such a situation, their achievements in generating domestic innovation capability show that the state, in this respect, has not been highly efficient. In fact, while the state and its

S&T policy are not solely to blame for the lack of industrial innovation in Brazil, they have contributed to it by sheltering domestic producers from foreign competition for too long a period, with an approach to S&T as a sector in itself, a centralization of S&T activities, a disarticulation of measures, and an instability of financial support to research activities. The point I want to make here is that these characteristics of S&T policy in Brazil are a result of the kind of insulation that characterized the policy-making apparatus. Because of its deficient 'embeddedness' with technology producers and users, state autonomy was more a liability than an asset for state capacity of efficient intervention.

Various attempts have been made to organize the S&T policy experience in Brazil in relatively homogeneous and distinctive phases.[7] For the purpose of analysing state capacity for S&T policy-making, I propose to organize these twenty-five years in four periods:

1 *1968–1973*: the repression years, when the economic miracle facilitated centralized planning and autonomy in S&T policy-making;
2 *1974–1984*: the political decompression years, when the end of the economic miracle and oil shocks led to disillusionment with planning and sectoral technology policies showed that Leviathan was also divided in technological matters;
3 *1985–1989*: the years of transition to democracy, when a civilian government forced to face an economic crisis, experimented with heterodox policies, and attempted to strengthen centralized S&T policy-making without a strategy for technological development; and
4 *the present phase* that started in 1990 with a democratically elected administration which introduced a new strategy for modernization within economic crisis and world recession, and has exposed the S&T infrastructure to destruction.

The first three phases are analysed in this sub-section and the last one, the current phase, is the subject of sub-section 2.4.

2.3.1 From 1968 to 1973

Governmental activity concentrated on institution building for S&T policy-making which involved the shaping of administrative structures and the defining of normative and financing instruments. S&T policy in Brazil was part of a conception of governmental administration based on modern planning techniques and on strategic considerations. It was expressed in the first National Development Plan (PND I, 1972–4) and the first Basic Plan for S&T Development (PBDCT I, 1973–4). In relation to its content, S&T policy supported the development of physical infrastructure and high-quality human resources for research in any area of knowledge and regulated the acquisition of foreign technology.

Administrative organization included the creation of new structures and agencies and the adaptation of existing bodies to new functions. In 1970, the National Institute of Industrial Property (INPI) was created to administer the technology transfer system, the patents and trademarks system, and the linkages between the domestic industrial research supply and demand. The existing relationship between the activities of CNPq (created as Research Council in 1952) and the planning system was formalized in 1972 through the creation of the National System for the Development of Science and Technology (SNDCT) to formulate and implement policy, with CNPq in charge of scientific and technological matters and the Ministry of Planning and BNDE in charge of financial matters. In 1972, the Industrial Technology Secretariat (STI) subordinated to the Ministry of Industry and Commerce was created to develop and implement industrial technology policy, and in the same year the National System of Science and Technology Information was set up as a network of data banks covering economic and social sectors; it was also during this period that institutional foundations of sectoral technology policies were established, escaping from the centralized S&T policy administration. This centrifugal movement, initiated earlier in relation to nuclear research, characterized not only the defence industrial sectors. It involved aeronautics, but also concentrated on agriculture, data-processing equipment and telecommunication.[8]

Two of the main normative instruments of S&T policy were issued in this period. One was the Brazilian Industrial Property Code (1972), granting protection except to pharmaceutical products, food, and the use of discovery of species of micro-organisms. The other was the Brazilian Copyright Law of 1973. Furthermore, the principal financing instruments of S&T policy in addition to FUNTEC were created: the National Science and Technology Development Fund (FNDCT) in 1969, and the Programme for the Support of Technological Development of the National Enterprise (ADTEN) in 1973.

A reform in higher education was initiated with the objectives of improving the quality of university teaching staff, stimulating scientific research, and providing adequate training to highly qualified human resources.[9] CAPES, an agency of the Ministry of Education, has been in charge of co-ordinating and operating the whole system. It also helps the Federal Education Council to issue authorization for the initial operation of each graduate programme. Scholarships are administered by CAPES and CNPq, and both these agencies, together with FINEP, have various financing instruments to support research and graduate programmes.

The state autonomous initiative of designing S&T policy goals was not complemented by the establishment of an administrative structure that could enhance policy cohesion. Instead, and according to a historical trend in Brazil, changes of the state apparatus for the accomplishment of its S&T policy tasks were made rather by addition than by transformation of previous structures. Recognizing the consequent segmentation of the administrative structure, an attempt was made at combining the various pieces into a National System for the Development of Science and Technology. At that time, however, this attempt involved only the various agencies within the Secretary of Planning. Besides the planning 'cluster' composed of CNPq, BNDE/FUNTEC, FINEP with FNDCT and ADTEN, and CAPRE (included in this group because of its role in the initial stages of definition of a sectoral technology policy), the other administrative pieces constituted two other clusters. One was the industrial cluster involving INPI and STI, and the other the education cluster composed of CAPES and the Federal Education Council. While the planning cluster had apparently greater power than the others, each one was relatively independent in setting its own policy goals. On the other hand, the clusters were not equally provided with the same level of bureaucratic expertise. The agencies in charge of S&T matters within each cluster were generally characterized by a relatively higher qualification than their administrative environment. Inter-cluster analysis shows that higher qualification in S&T matters was concentrated in agencies of the planning cluster. This happened especially because of BNDE's widely recognized technical and professional integrity (Willis 1986), which later extended to FINEP who absorbed BNDE's procedures and some of its staff. CNPq had since its creation been strengthened by the contribution of respected scientists whose activity in the peer review groups has enhanced the respectability of decisions taken. It was, however, more vulnerable to patronage than other components of the planning cluster. Recruitment for the S&T state bureaucracy was dominantly meritocratic in BNDE and FINEP, mixed with political appointments in CAPES, STI, INPI and CNPq, and dominantly political in the Ministry of Education. Even meritocratic procedures were not in accordance with fully universalistic criteria. Despite the admission to public service based on a solid curriculum vitae, personal connections dominated the way of access to job opportunities. The last BNDE public competitive recruitment examination was held just before the military coup (Willis 1986).

It was in relation to the articulation with societal groups that the state showed its strong limitations during this period. Relationships with local industrialists was deficient during military rule. In the presence of weak and forcefully reformed political parties and a puppet Congress, the only institutionalized channel of political representation was the old corporatist structure. However, against all odds, there was some improvement in the organizational capacity of Brazilian industrialists, particularly those of the most dynamic sectors, who created specialized organizations to aggregate and articulate their interests. These organizations eventually constituted a 'parallel' structure to the corporatist form of representation (Stepan 1985; Diniz and Boschi 1989). The local bourgeoisie had a strong affinity with the regime, legitimate in its eyes, but they have never been connected by any well-institutionalized system of linkages (Martins 1986; Evans 1989). 'Bureaucratic rings' existed and linked some individual businessmen to particular state officials within the bureaucracy. These links were, however, ad hoc and informal.

Institutionalized relations between the state and scientists and engineers, the scientific and technological community, were equally nonexistent, but in contrast to the industrialists, there was no affinity involved in this case and the relationship was highly conflictive. Members of the scientific community had been politically active during the Goulart administration and, together with university students, took part in the mobilization in favour of 'basic reforms' that threatened the propertied classes. The repression that followed the coup also affected university students and the scientific community, some of the most eminent members of which were punished, imprisoned, and ended by leaving the country.[10] In 1967, the 19th annual meeting of SBPC (the Brazilian Society for the Advance of Science) signalled the existence of some room for conciliation. SBPC's new President, W. Kerr – who had been briefly imprisoned during the repression of three years before – sided with the government with regard to the nuclear energy policy and called for greater expenditure in research. Criticism against the reform of university education and the repression of scientists was strong. Discontented students took to the streets and together with guerrilla movements, and backed by the electoral results that favoured opposition, they triggered the emergence of the period of terror. The launching of S&T policy during this period, and particularly its 'science' bias, indicates that together with instrumental goals it also played a political role. It is important to note that scientists did not participate in the elaboration of the S&T chapter of the Strategic Development Programme and CNPq's draft was overshadowed by the Ministry of Planning and General Co-ordination (Botelho 1989). University education reforms, justified by students' and their middle-class families' demands for increase in college enrolment, opened the system to private initiative. This would later prove to have contributed to the heterogeneity in standards of college training.[11] The credit system (of disciplines) replaced the traditional class composition, thus contributing to the political demobilization of university students. Also, institutionalization of graduate programmes, with available resources for research and scholarships for study in the country and abroad, resulted in the co-optation of university researchers. More than that, allocation of resources for graduate programmes and research groups was centralized in the Ministry of Planning and, therefore, beyond the reach of university power structures. The result of these reforms was the emergence of a new system of power relations within the social space of science (Botelho 1989). While the co-optation of university researchers gained momentum, repression continued to cause the dismissal of various professors and, in what is known as the 'slaughter of Manguinhos', the majority of researchers of this biology research institute in Rio de Janeiro lost their jobs. From 1972 until the end of military rule, the annual meetings of SBPC became the 'dose of oxygen to the freedom of expression of students and intellectuals' (Botelho 1989).

2.3.2 From 1974 to 1984

This period witnessed the climax and decline of planning as an administrative tool in Brazil. The second and third national development plans (PND II, 1974–7, and PND III, 1978–81) epitomize those two movements: the carefully defined and integrated quantitative targets of the second were substituted by general statements of intentions of the third. The Brazilian enthusiasm for planning was transformed into disappointment by the two rather unpredictable oil shocks and a change in style of the administration. The PND II put strong emphasis on the energy sector and industrial technology. The development of basic industries with high technological content, such as computers, capital goods, petrochemicals, steel and aeronautics was made a priority. The PND III established the general goal of reducing dependence on science and technology. The second and third Basic Plans for S&T Development (PBDCT II, 1975–9; PBDCT III, 1980–5) emphasized co-ordination, but remained practically unknown and mostly ineffective. While the scope of S&T policy in the preceding phase was dominated by 'scientific' bias expressed in various means of support for human resources training in research in all areas of knowledge, now the technology component surfaced and became a priority.[12] New programmes were created to stimulate technology absorption and activity at firm level and to bridge research centres and production.[13] The first and second National Plan for Graduate Studies (PNPG I, 1975–9; PNPG II, 1982–5) institutionalized graduate studies as a step towards an academic career; it supported training of high-quality human resources for research and teaching; and promoted significant improvement in the quality of graduate programmes in Brazil. Agencies in charge of S&T policy became the major source of support for training of human resources for research in the universities and research centres, while other agencies such as BNDE turned to support technology activities at firm level.

Administrative reform strengthened the power of planning agencies. At the centre of such reforms, the Ministry for Planning became the Planning Secretariat (SEPLAN), acquiring broader powers and attached to the presidency. CNPq became he National Council of Scientific and Technological Development and, placed under SEPLAN, was the central agency for planning, co-ordinating and implementing science and technology policy.[14] However, the issue of how to harmonize administrative competence over technology policy remained untouched in spite of the strengthening of central agencies for technology policy-making.[15] This was evident by the fact that sectoral technology policies continued, following their own unco-ordinated courses and in the technology policy role of the Ministry of Industry and Commerce.[16] In relation to sectoral technology policies, two points deserve mention. One was the consolidation of the defence industry with the creation in 1975 of Nuclebras (Brazilian Nuclear Enterprises SA), partner with CNEN in R&D activities. The other was the creation in 1980 of the Special Secretariat for Informatics (SEI), under the National Security Council to substitute for CAPRE.[17]

Among the normative policy instruments issued in this period, the most important were the Policy of Equipment Acquisition and Technological Development of Telecommunications (1978), the National Programme for Biotechnology, PRONAB (1982), and the Informatics Law (1984). This indicates the level of articulation of sectoral S&T policies. Using the procurement power of the state, Telebras promoted the modernization of Brazilian telecommunications and local technological development. In 1981, contracts for new telephone exchange equipment with stored programme control were restricted to the CPA-T type, manufactured within the country by Telebras's R&D institute, CPqD. PRONAB proposed actions for training human resources for research, disseminating new technology, and encouraging production in the sectors of agriculture, energy and health, with significant participation of private industry. In 1984, the Informatics Law institutionalized the reservation of the local market for locally owned IT companies.

Financing instruments were strengthened by the creation in 1974 of a new subprogramme under BNDE/FUNTEC to provide incentives for R&D projects at firm level, and in 1976 by the creation of the Scientific and Technological Research Incentives Fund (FIPEC) within the Bank of Brazil. Three additional institutions were created within the BNDE jurisdiction: the Brazilian Investments, Inc. (IBRASA) to invest in Brazilian enterprises; the Brazilian Mechanics, Inc. (EMBRAMEC), to provide capital for national capital ventures in the capital goods sector; and the Basic Goods, Inc. (FIBASE), to provide capital for basic goods.

Modernization of S&T state apparatus by transformation of older structures had an important implication for policy coherence. CNPq, which had been since its creation in 1951 in charge of granting scholarships and research funds to individual researchers, and had, consequently, dealt with issues affecting science policy, was promoted to the role of co-ordinating the whole S&T system. The agency lacked not only the competence and administrative experience but also the legitimacy within the technology system to achieve this task. It also became highly vulnerable to the patronage that accompanied the political decompression, particularly during the last military government. Although the expertise of the central planning agency was challenged by tasks beyond its administrative culture and although its numbers were threatened by political appointments, the various sectoral 'pieces' of the technology policy apparatus strengthened the expertise during this period and enhanced their corporate culture for the promotion of technology development. At the end of this period, S&T policy expertise was concentrated not only in the technical agencies of the planning cluster but also in sectoral organizations in the petrochemical, defence, communication and computer areas. These sectoral clusters became known as 'parallel' S&T policy systems (Souza Paula 1991; Conca 1992). While S&T policy became increasingly fragmented, 'pockets of efficiency' emerged.

Political decompression did not imply significant change in the mechanisms of representation of interests. On the contrary, by 1975 the decision-making process was much more closed and the industrialists' representative associations were progressively excluded from strategic decisions taken on general goals of economic policy. The dual structure of representation continued to grow and in some high-tech industrial areas such as informatics they became highly influential. An 'antistatist' campaign in the mid-1970s displayed the disagreement of Brazilian businessmen with the regulatory expansion of the state. By 1978 with the issuing of the 'Manifesto of the Eight' entrepreneurs, a significant fraction of the state's initial allies had joined the civil movement for liberalization (Stepan

1985). By the early 1980s, it became clear that the Brazilian state had lost much of its capacity to lead its allies. Links with societal groups remained informal and ad hoc. Representatives of private groups were included in the composition of *conselhos* (councils), the high-level, decision-making bodies closely connected to the office of the presidency. This representation provided the regime with information, but apparently private groups were not able to use the *conselhos* more than the 'bureaucratic rings' as effective mechanisms for venting their demands (Geddes 1986; Cerqueira and Boschi 1977). However, a closer relationship between state S&T agencies and local industrialists, domestic and foreign, was achieved in specific areas such as the petrochemical, defence equipment, aeronautics, telecommunication and computers. In these areas, concentrated expertise resulted in well-designed and coherent programmes while closer collaboration with industrialists improved efficiency in implementation.

In many other fields, decisions in S&T matters were exclusively centralized. In this respect, innovation in this period consisted of the decision to submit the informatics policy to Congress for discussion. The Informatics Law was a rare case in which S&T matters went beyond the Executive settlement. Similar to the position of businessmen, participation of members of the scientific community in some spheres of the S&T policy apparatus did not mean that they had any substantial role in decision-making, nor that their demands and/or criticism were taken into account. Scientists, in any case, did not go further than acting as consultants in the assessment of research projects and in the granting of scholarships or in the assessment of the state-of-the-art of their areas of knowledge in connection with follow-ups to basic plans of scientific and technological development. However, most of the top positions at CNPq were occupied by former scientists or people experienced in university administration. While this fact does not necessarily entail the 'capture' of this agency by corporatist interests, it may signal the limitations of this central planning institution in dealing with broader S&T issues and penetrating various other groups as demanded by its function of co-ordinating activities.

2.3.3 From 1985 to 1989

In this period the civilian government did not do much in terms of S&T policy. Apparently the Sarney administration had no specific project for S&T besides the creation of the Ministry for Science and Technology (MCT) and budget increase. With the creation of MCT, CNPq lost its formal position as co-ordinating agency and returned to its original role of supporting individual researchers according to their areas of knowledge.

The new ministerial status seems to have added very little to S&T policy. CCT, the central decision-making body, was convened only once in five years (SCT/PR 1991: 5). The ministry, however, made an attempt at widening S&T policy into technological issues. The Training Programme of Human Resources for Strategic Areas (RHAE) was created to cater to private enterprise demand and four new departments were created within MCT to complement the ministry's supervision of new technology developments, in addition to the Special Secretariat for Informatics (SEI) that had been incorporated in the ministry. The new departments were the secretariats for biotechnology, advanced materials, fine chemistry and precision mechanics.

The normative activity of technology policy in the period concentrated on regulation in the area of information technology. The Brazilian 301 case – the conflict with the US on Brazil's informatics policy – challenged the administration's negotiation capability and made clear the disarticulation among policies and within the various state agencies in relation to technology policy. Eventually, the conflict weakened SEI, created political commitments to ending the protectionist policy and, in the subsequent administrative reform of 1989, MCT was demoted to Secretary of Science and Technology (SCT). The launching of the First Plan for Informatics and Automation (PLANIN I) was not substantially affected by the conflict with the US, but the preparation of the Software Bill and the approval of the Software Law were made under American monitoring. Resources for FNDCT were nominally enlarged in the beginning of the period, barely preventing it from total collapse.

The political pact that presided at the transition to democracy left little room for the creation of stronger democratic institutions that could serve as efficient mechanisms of representation of interests. Programmatic parties were not created, perhaps with the exception of the already strong left-wing PT (Workers' Party) and, on a minor scale, the centre-left initiative of PSDB (Social Democracy Party). None of them had proposals for the S&T area. The Congress, which added the task of preparing a new Constitution for the country to its regular duties, had been elected under the same electoral rules that benefited the smaller and poorer regions, the stronghold of traditional elites.[18] The new Constitution of 1988 strengthened the role of the legislature and the republic (whose component states received a larger share of the state budget without, however, the corresponding duties) and allocated a certain proportion of the state budget to education. It also contained a new chapter devoted to science and technology.

The creation of MCT by the New Republic was a response to an old demand of a significant part of the scientific community (Schwartzman

1989). Soon after the new government took office, representatives of the scientific societies voted the creation of a permanent commission to follow up the organization of MCT, to discuss the reformulation of the agencies put under the new ministry's jurisdiction, and to influence solutions to the problems affecting the S&T area. According to this commission, the democratic regime did not substantially alter S&T policy inefficiency inherited from the military regime. MCT was never a strong institution. It was politically marginal, vulnerable to state's patronage, and impotent to protect research activities from the effects of economic crisis. The New Republic had no S&T policy and its agencies and research institutes had to compete with interest groups for public resources. By the end of the 1980s, the scientific community demanded urgent measures be adopted to avoid total collapse of research institutes, the preparation of a bill defining criteria for allocation of resources that could bring financial stability to S&T activities, and an institutional location for the policy that could restore its centrality and relevance (Schwartzman 1989).

Also, links with the scientific community during the transition government did not significantly change from the situation that prevailed in the last military governments. Nominally, at least, they gained access to decision-making bodies such as CNPq's Decision-Making Council and MCT's council. In addition, the process of recruitment to these bodies changed from the previous selection by the government to proposals made by the scientific societies and institutions. These changes had, however, very small practical implications because the responsibility of decision was not really transferred to these formal decision-making bodies (Schwartzman 1989).

S&T policy would have to contain, from the standpoint of the scientific community:

1 a political commitment to enhance the contribution of S&T to development translated into long-term perspective and stability in research investment, together with the creation of instruments to facilitate the incorporation of research results into production;
2 urgent measures for the provision of adequate operating conditions for research institutes;
3 an increase in the S&T budget, with a significant part under the control of the planning agencies, to be effectively spent in research activities;
4 improvement in administrative efficiency of the agencies; and
5 more specific measures such as the creation of state-level agencies for funding research, exemptions for research institutes from import

restrictions on equipment, and an assessment of the existing programmes of institutional support (Schwartzman 1989).

While the transitional government did not react to the scientific community's demands, they remained a matter to be considered by a fully democratic administration.

2.4 Collor's project of conservative modernization

The first democratically elected President in Brazil after military rule, Collor de Melo, found very favourable conditions for policy change. There was – and still is – a consensus on the need for drastic measures to control inflation, and a wide awareness regarding the exhaustion of the import-substituting strategy of industrialization, which weakened possible resistance to change. These exceptional conditions gave the administration the necessary degree of autonomy to define and implement a stabilization plan and to design a project of structural change. Measures derived from orthodox economic thinking and prescriptions were implemented to stabilize the economy, albeit some of them by rather illiberal means. The project of structural change was a composition of elements taken from the liberal, ecological and social democratic agendas, thereby satisfying demands of a wide political spectrum.

A significant institutional change was planned to put firms at the centre of the effort of S&T development, stressing that they would have to play a much more active role in technological development than under the previous protectionist strategy. The main target set out in the government's General Directives for Industrial and Foreign Trade Policy (PICE), of 1990, was to increase efficiency in the production and commercialization of goods and services through modernization and the restructuring of the industrial sector.

Collor's Programme for Industrial Competitiveness approached the factors determining competitiveness at three levels. At structural level, it aimed at stimulating private investment, promoting exports, reforming technical and higher education, and complementing the programmes of productivity and quality of industrial firms (PBQP and PACT) by deregulating technology transfer. At sectoral level, emphasis would be given to those with comparative advantages already defined by their export achievements and to those that could generate and diffuse innovation and technical progress. Technological, production and organizational modernization of sectors with comparative advantages already defined would be stimulated, and an association of locally owned companies with foreign capital fostered in high-tech sectors. At firm level, the project promoted the restructuring of the Brazilian entrepreneurial model, through mergers, de-verticalization and privatization.

The Brazilian Programme of Quality and Productivity (PBQP) was launched to support the modernization effort with horizontal and sectoral sub-programmes. Horizontal measures taken for improvement of quality and productivity would include mobilization of industrialists and workers for improvement of quality and productivity; development and diffusion of managerial methods; development of human resources; adjustment of technology services; and institutional articulation of government, industry, commerce, service companies, education institutions, and science and technology institutes. Sectoral programmes of quality and productivity (PSQP) would be formulated for some sectors of the Brazilian economy to be identified, and also for segments of the Public Administration at national as well as at state level. Although data on the implementation of PICE still have to be produced, measures taken along its lines were particularly those related to the elimination of protectionism.

In 1991, a programme for Support of Technological Capability of the Brazilian Industry (PACT) was launched. This programme targeted technological capability in mature as well as new technologies industries. PACT defined actions to be taken along two lines. One would be the creation and strengthening of the conditions external to the firms, such as development of human resources, motivation and mobilization of entrepreneurs, modernization of technological institutes and research centres in public and private universities, and strengthening of the network of technology information. The other line of action would be to give direct support to firms for R&D activities in general, strategic programmes of development of new product or process, purchase and absorption of technology, development of stronger links with research institutes and universities in technology parks, and marketing.

Collor's project of structural change can be summarized in three major points:

1 elimination of all protectionist barriers and similar means perceived as obstacles to the free play of market forces;
2 destruction of the entrepreneurial role of the state and its substitution by a government strengthened in its decision-making system;
3 modernization of production to increase international competitiveness of Brazilian exports.

What societal forces support Collor's project? After the initial paralysis provoked by the unanimously supported stabilization plan, and particularly because of its disappointing results, social forces are starting to realign and this movement will help identify the supporters of the modernization project. The elimination of all protectionist barriers, once welcomed by many, is now provoking obvious reactions because it is affecting groups of industrialists very differently.[19] A few of the bigger domestic groups will be able to resist the structural change and the economic crisis. The project seems to imply, therefore, an abandonment of whatever intention there was of building 'national capitalism' in Brazil (Martins 1990). Foreign capitalists are likely to be the beneficiaries of the 'gale of creative destruction' which will result whenever there is a forced push to make domestic industrialists face competition (Evans 1989). Of the triple alliance who supported the project of 'conservative modernization' under military rule, the only recognizable partner still actively in favour of this new version of modernization is the foreign capitalists. State entrepreneurs are certainly in a very uncomfortable position and many of the domestic industrialists are being overtly antagonized. Traditional elites, whose political privileges were preserved in the transition to democracy, have strongly supported the project in a patrimonialist exchange. However, as signalled by the political mobilization that resulted in the President's impeachment, the other forces in support of the modernization project have agreed to set a limit for the action of the traditional elites and their predatory behaviour.

The Collor administration's project of deregulation, economic liberalization and technology modernization was to be implemented with as little state intervention as possible. In the administrative reform towards a minimalist state, the Secretary of Science and Technology (SCT), one among other agencies, was not granted any special link with central decision-making. CNPq and FINEP saw their financing resources to support researchers and research centres vanish as a result of the financial crisis of the state and the political use of *contingenciamento* (liberation of funds depending on revenue).[20]

S&T policy bureaucrats of central planning agencies and sectoral institutions were discouraged by salary cuts and demoralized by the administrative arrogance of the new team who treated the agencies as if they all were part of the traditional or 'spoils' bureaucracy. Many of the most qualified left the state bureaucracy to start (or continue) academic careers. Others that remained in their positions have not had much to do to boost their morale and restore motivation. Those who had contributed to the formulation of the new strategy were disaffected with the first public signs of corruption.

The autonomy the administration enjoyed was not followed by actions towards building social support necessary to the project of structural change; nor was any significant move made towards building stronger democratic institutions. In this respect, the impeachment of the President did more for Brazilian democratic institutions than any of his acts during three years in power. The administration did not show any interest in building a permanent parliamentary majority and rather demonstrated a preference for decisions made oblivious of the Congress and political parties. Popular mobilization was pursued directly through the media, the 'tv-populism', one of the expressions of the President's 'political Calvinism' (Schneider 1991a). Businessmen were pushed to negotiate individually with the government, particularly for the return of frozen financial assets. The corporatist representation in sectoral 'chambers' was a slight variation of the mechanism used during military rule. The Executive Groups for Sectoral Policy (GEPS) were created to link the private sector's demand and the Ministry of Economy in matters of taxation, exchange and tariff policies, with the objective of channelling these demands to one single counter and facilitating policy co-ordination. However, despite being taken as a new method of industrial policy-making, GEPS have been kept as an informal arrangement. These corporatist representations of sectoral interests in governmental bodies may have risky implications. It may generate mechanisms of private appropriation of public resources, subvert the basic principle that managerial decisions should be taken at firm level, and make it difficult to phase out state support when required (Fritsch 1990: 347).

The scientific community, one eminent member of which had been chosen to occupy the Secretary of Science and Technology, was pushed into an opposing stance by the pragmatic approach to modernization and the persistent lack of resources for research. In addition, university researchers were systematically accused of inefficiency within the wave of criticism of the state bureaucracy and state-supported organizations. The governmental orientation towards science and scientific research was made clear in the first months of the new adminstration:

> in a country where investment in S&T is so low, the allocation of 90 per cent to pure science, I mean, to research without immediate use in industry, is a scandal . . . Priority will be given to research that can be used in the production process. Effective demand for research by industrial companies will set this priority.
>
> (Luis Paulo Veloso Lucas, Director of Industry and Commerce, Ministry of Economy. Quoted in Mundim and Nelson 1990, my translation)

A reaction of the scientific community was presented in the subsequent annual meeting of SBPC and echoed within the most moderate and conservative newspapers. 'That the university should produce research that serves the needs of the industrial sector is almost a consensus among the various currents of opinion within the academia. To direct all research towards this end is, however, dangerous' (O Estado de São Paulo 1990, my translation). Within the government, however, the criticism was received as lack of support and, therefore, was unacceptable. The scientists were accused of corporatism and the SBPC, whose positive role in the defence of freedom during military rule was acknowledged, was blamed for being transformed into a political opposition group (Goldemberg 1990). Conflict between the government and the scientific community gained wide press coverage during the three years in which Collor was kept in power. The central point of the dispute was the withdrawal of state support to university research which, in turn, would have to be dependent on private demand for the necessary funds. For the scientists, the government was 'liquidating' science, and the Minister of Economy was the 'exterminator' of all (Nussensweig 1992a, 1992b).

It was not only criticism that came from the ranks of the scientific community. The contribution of the university for industrial modernization was seen by some as a historical event because, for the first time, the academy would be acting jointly with industries 'in the light of something close to a project of national development' (Vogt 1990, my translation). Technological researchers were, in principle, supportive but in practice their enthusiasm was thwarted by the prospect of discontinuity of long duration projects. This affected in particular, high-tech research institutes (Medeiros 1990), most of which were state sponsored and, therefore, at the risk of losing research funds. Some others were involved in defence-related projects whose continuity had been under strong attack by foreign organizations and countries (Jornal de Brasília 1991a, 1991b).

Criticism was also made by industrialists who had supported the modernization project. One set of such criticism was made by industrialists who were enthusiasts of complete deregulation and trade liberalization. Moved by their extreme liberalism, they still complained that changes were not drastic enough; that what was left of the informatics policy was still unacceptable; that no barriers of whatever nature should remain for the importation of equipments inputs, components and even raw materials; and that an interventionist or statist bias was still found in the governmental proposal of increasing resources for S&T (Folha de São Paulo 1990; Nakatani 1991). Other criticism, less radical, demanded the creation of incentives for investment in R&D and was supportive of stronger technological collaboration between industry and university research institutes (Musa 1991).

Technological partnership with university research centres was not an initiative that started with the current modernization project. The link between research and industry, that had been attempted by the NAIs much earlier, was pushed in the 1980s by the creation of S&T 'poles' or technological parks, particularly in high-tech areas. In the beginning of the 1990s, there were twelve such high-tech poles in which a strong interaction between industry and university researchers was being developed, particularly through shared utilization of research facilities. Another illustration of previous experience is the Multidisciplinary Research Centre on Chemistry, Biology and Agriculture (CPQBA), created within Unicamp in 1987, and which had established productive links with companies of the chemical and pharmaceutical sectors. The decline in resources for research funding on the one hand, and the pressure for modernization in a juncture of economic crisis on the other, have favoured the convergence of research centres and industrial companies.

The modernization project stimulated these partnership initiatives. The Office for Technology Transfer of Unicamp, created in October 1990, had just one year later a portfolio of around 300 research outputs to be transferred and, in its first eight months, was visited by an average of 40 businessmen monthly (Vogt 1991). There is no information on the transfers effectively made. Early in 1992, other initiatives were announced. Uniemp (university-enterprise) – one of these initiatives – is to be a body promoting the link between research and production. Initially, it will involve ten to fifteen colleges, and various industrial companies among which are the domestic Metal Leve, and the local subsidiaries of Rhône-Poulenc and Mercedes Benz. Uniemp activities will be financed with the resources of a Fund for Technology, to be created in the state of São Paulo, composed of 0.3 to 0.5 per cent of the state tax revenue, donations and funds from international agencies. While these initiatives and the development of technological parks as well as their extension to mature technologies may be signalling positive S&T effects of creative destruction, the withdrawal of state support to science and other research activities in the Brazilian universities may be pointing to a far-reaching negative effect of the crisis. Conditions for interaction between university research and industry 'happened not because such interaction was established by law or policy, but because of previous accumulation and concentration of S&T infrastructure' (Medeiros 1990).

The first civilian alternative policy project for technological development has been hampered by economic and political obstacles. Economic and political restraints emerge from the timing of measures, the subordination of industrial and technology policies to the programme of economic stabilization, indiscriminate shrinking of state intervention, and the financial crisis of the state. Timing and different intensity of measures for opening the economy in a situation of domestic economic crisis and global recession on the one side, and promotion of industrial modernization on the other, have exposed the industrial structure to strong pressure. Public resources for S&T were subject in fact to reductions, instead of the proposed increases. If the concern with technological advance of the exporting sector means more than improving production capacity, then an adequate S&T infrastructure, a strong educational system, and some form of promotion of innovative domestic industry is required. Measures towards a 'lean' state bureaucracy, cuts in funds for S&T, weakening of the university system, and an indiscriminate opening of the industrial sector may hinder, instead of promote, the innovation capacity of industries in Brazil (Suzigan 1991). Technological development in Brazil requires a stronger, more qualified, better paid and more engaged bureaucracy, able to design and adapt coherent policies. Moreover, stronger, stable and institutionalized mechanisms of representation of interests are necessary for the state to consolidate democracy and improve its autonomous capacity for policy design and implementation in creative co-operation with societal groups.

3 CONCLUSION

Brazilian political institutions have not developed to a point where they can allow autonomy in economic policy-making and at the same time provide organized and stable mechanisms of interest representation (Hagopian 1990; Martins 1990). Relationships between decision-makers and societal groups were made either in a patrimonialist way – before and during military rule – or were simply nonexistent. Autonomy in economic policy-making in Brazil has often had a strong component of despotism, be it civilian or military. While policy design has been more dependent on bureaucratic expertise, lack of proper links with societal groups has contributed to reduced state capacity for S&T policy implementation.

NOTES

1 The following persons were interviewed in July 1992 for the present study: José Paulo Silveira and Manuel Lousada Soares (Technology Department, Secretary of Science and Technology), Adolpho Wanderley da Fonseca Anciães and Eduardo Baumgratz Viotti (Planning Department, Secretary of Science and Technology), Maria Carlota de Souza Paula (Advisor, House of Representatives), Carlos Alberto Moreira Filho (Department of Microbiology, University of São Paulo), Edson Fermann (Technology Department, Federation of São Paulo Industries), Luciano Coutinho (University of Campinas), Mauro Fernando Maria Arruda (Institute for Studies of Industrial Development, São

Paulo), Antônio Fernando C. Infantosi and A. Claudio Habert (Graduate Programmes in Engineering, Federal University of Rio de Janeiro), Carlos Alberto A. Carvalho Filho (Catholic University of Rio de Janeiro), Lourival Carmo Mônaco and Wilson Chagas de Araújo (Studies and Projects Financing Agency – FINEP), Carlos Santos Amorim Junior, Alvaro Rodrigues dos Santos and Francisco Frederico Sparenberg Oliveira (Executive Group for Corporate Modernization, Secretary for Science and Technology of the State of São Paulo), José Israel Vargas (Brazilian Academy of Sciences), Lindolpho Carvalho Dias, Getúlio Valverde de Lacerda, Manoel Marcos Formiga, Guilherme Brandão, João Augusto Bastos, Celina Roitman and Itiro Iida (National Council for Science and Technology Development – CNPq), Marisa Cassim and Fernando Spagnolo (Commission for the Improvement of Tertiary School Teachers – CAPES).

2 For recent analyses of the Brazilian S&T system, see: Schwartzman 1989; Vaitsos 1990; Suzigan 1991; Coutinho 1991. Recent discussion of competitiveness of Brazilian industrial exports and the capacity of industrial firms to innovate in Brazil is found in: Fajnzylber 1990, 1991; Schmitz and Cassiolato 1992; Carvalho 1992; Ferraz 1992; Ferraz, Rush and Miles 1992; Meyer-Stamer 1993. For an analysis of the Brazilian education system and educational constraints to innovation, see: Castro 1984; Barreto, Gusso and Demo 1990; Durham and Gusso 1991; IEDI 1992; Villaschi 1992.

3 Capacity does not have a special meaning and is used here, as in the literature on stabilization policies, to denote 'ability to'. It does not imply in itself any idea of the intensity or quality of this ability nor of the quality of the policy that results in the use of this ability.

4 See Mowery (1994) for a useful discussion of the concept in the context of industrial countries.

5 Highlights of this previous activity include the creation of the University of São Paulo and of the National Institute of Technology (INT) within the Ministry of Industry, Labour and Commerce, in 1934; the foundation of the Military Technical School, later to become the Military Engineering Institute (IME), in 1935; the creation of the National Research Council, later to become the National Council for Science and Technology Development (CNPq) and the Campaign for the Improvement of Higher Education Personnel (CAPES), in 1951; the foundation of the National Development Bank, later to become the National Bank for Economic and Social Development (BNDES) to finance long-term development projects, in 1952; and the creation of the Radioactive Research Institute, later to become the Centre for the Development of Nuclear Technology (CDTN), in 1953. Before formally defining technology policy as an explicit target, the military government had created, in 1964, the Scientific and Technological Development Fund (FUNTEC) and the Special Agency for Industrial Investment, both within BNDE, and had given CNPq the task of programming science and technology activities. In 1967, FINEP, a financing programme of BNDE for supporting the technical design of industrial and agricultural projects, was transformed into an independent public enterprise, and the Fund for the Support of Technology (FUNAT) was created within the Ministry of Industry and Commerce to train industrial technicians.

6 The following discussion on alternative explanations for the origin of S&T policy in Brazil is based on Erber (1979).

7 Bornstein (1988), Adler (1987) and Coutinho (1991) have different classifications. Adler takes the period prior to 1969 as one of 'technological *laissez-faire*' and organizes the explicit S&T policy from 1969 to 1982 into three phases: 'creative' (1969–75), 'stabilization' (1976–8), and 'drought' (1979–82). Bornstein considers 'implicit' the policy before 1960, and classifies up to 1979 in three phases: 'military coup and stabilization' (1964–7), 'initial formulation' (1968–74), and 'industrial technology' (1975–9). Coutinho classifies S&T institutional evolution into four phases marked by the creation of CNPq (1951–64), the creation of FUNTEC (1965–74), the transformation of CNPq into the central S&T policy agency (1974–84), and the approval of the Informatics Law (1984–90).

8 Brazilian 'informatics' policy originated from the Special Working Group that was set up in 1972, with the support of FUNTEC, to study the basis for domestic design and production of data-processing equipment. In the same year two agencies with a central role in the later formulation of disarticulated policies for Brazil's electronic complex were created. Under the jurisdiction of the Ministry of Planning, CAPRE later became the decision-making centre of the informatics policy. Under the Ministry of Communication, TELEBRAS later was in charge of defining criteria for state procurement of communication equipment, mostly based on digital technology. In the area of nuclear policy there were the following developments: an agreement for scientific and technological co-operation with Germany was signed; CNEN, that had taken over the normative and R&D functions from CNPq since 1956, was moved from direct subordination to the presidency to the jurisdiction of the Ministry of Energy; decisions were made in favour of light-water reactors over the academic proposals of natural uranium ones; and a Westinghouse plant was chosen. The policy for technology capability building in aeronautics, decided within military circles and in close relationship with a strategy for the Brazilian defence industry, was marked by the incorporation, in 1969, of EMBRAER, the Brazilian Aircraft Company to supply civilian and military markets with products designed at the Aerospace Technology Centre (CTA).

9 A system of incentives and controls was established consisting of (a) a requirement for graduate degrees for promotion in a university teaching career; (b) scholarships for graduate studies in Brazilian academic institutions; (c) scholarships for graduate studies in well-known foreign academic institutions; (d) state support for infrastructure and operational costs of, as well as research grants for Brazilian graduate programmes; (e) authorization for the operation of graduate courses conditional on positive assessment of their structure, qualification of teaching staff, and scientific production; and (f) periodic assessment of structure and achievements of each programme.

10 Among the thousands of scientists and university researchers punished because of their political views were: in the University of São Paulo, Mario Schenberg, renowned theoretical physicist, one of the founders of the university's Physics Department; Isaias Raw, a leading biochemist and the director of a foundation for science education; Luis Hildebrando Pereira da Silva, parasitologist; Julio Puddles, biochemist; Fernando Henrique Cardoso, sociologist. The former President of the Brazilian Society for the Progress of Science (SBPC), W. Kerr, was imprisoned for a brief period; various federal universities had their campus or premises invaded; and the respected educator Anisio Teixeira was dismissed

as Rector of the University of Brasilia, together with economist Celso Furtado and many others who formed the country's scientific elite. In October 1965, a politically based crisis at the University of Brasilia led to the resignation of its Rector and the dismissal of 90 per cent of its faculty. Leite Lopes, a physicist and former Vice-Director of SBPC had to resign from his function at the Brazilian Centre for Physics Research (CBPF) and decided to leave the country. Between July 1964 and June 1965 about one hundred scientists, medical doctors and engineers emigrated to the United States (Botelho 1989: 52). A larger number would, in later periods, leave the country for Europe.

11 Some analysts suggest that heterogeneity in standards and concentration in subjects have been consequences of the 'liberalization' of higher education in Brazil which became affected by considerations of increasing profits and lower operational costs (Schwartzman 1979; Villaschi 1992).

12 The prominence given to technology policy is also expressed in human resources training. Simultaneous with the National Plan for Graduate Studies (1974) was also created the Programme of Research Administrators (PROTAP) within FINEP for the strengthening of human resources in national enterprises.

13 Nuclei of Articulation with Industry (NAI) were created in 1975 for the benefit of public enterprises. As an instrument of 'buy national capital goods', their role was one of promoting the purchase and development of national manufactured equipment. In the same year the Co-ordinating Commission of Articulation with Industry (CCNAI) was created under FINEP. In 1979 the Industrial Technology Company was created within INT to transfer the technologies produced by the institute to the productive system.

14 The restructuring of SNDCT also included the creation of sectoral organisms called 'Technological Secretariats' in every ministry, all integrated to the SNDCT, and co-ordinated by CNPq.

15 The policy-making, co-ordinating role of CNPq was to be exerted by its Scientific and Technological Council (CCT), composed of representatives of various ministries. This Council has never had effective co-ordinating power.

16 The Ministry of Industry and Commerce, through INPI, kept the power to regulate technology transfer and to make decisions central to the import-substituting strategy. In 1975, INPI issued a normative act dividing technical know-how into five categories and requiring separate contracts or agreements on each so as to open the 'technological package'. In 1978, INPI barred payments by automotive subsidiaries to their parent companies for R&D on new models. In 1974, the National Council for Metrology, Norms and Industrial Quality (CONMETRO) became the National Institute for Metrology, Standardization and Industrial Quality (INMETRO) and was placed under STI jurisdiction within the Ministry of Industry and Commerce. This institute would later be placed under the jurisdiction of the Ministry of Science and Technology (MCT) in 1985.

17 Autonomy of sectoral technology policies can be illustrated by the cases of agriculture and nuclear research. Agriculture research centres, the training of human resources for research, and the bridging of research and production, were all co-ordinated and implemented by the Ministry of Agriculture with complete autonomy in relation to explicit S&T agencies. In the case of nuclear research, Nuclebras constituted a close-knit and highly insulated network with CNEN, including the Dosimetry Institute, the Nuclear Information Centre, the

Energy and Nuclear Research Institute, the Centre for the Development of Nuclear Technology and the Nuclear Engineering Institute. Other research centres involved: Aerospace Technology Centre, the Military Engineering Institute, and the Co-ordination of Graduate Programmes in Engineering. Nuclebras established in the early 1970s the Human Resources Programme for the Nuclear Sector (PRONUCLEAR) to develop human resources for the implementation of the 1975 Brazil–West Germany agreement, responsible for training more than 2,400 engineers and technicians in a seven-year period (Adler 1987: 309). In fact, there were two separate nuclear programmes: the 'civilian', in co-operation with Germany and relatively open to public scrutiny, and the Navy or 'parallel' programme, kept under secrecy until the end of military rule.

18 Because of these rules, the Northern/Northeastern regions, with 36 per cent of the Brazilian population and 18 per cent share of GDP, have 264 representatives in the Congress (House and Senate together). The Southeast/South, with 57 per cent of the population and 76 per cent share of GDP, have 267 representatives in both houses.

19 One example is the automobile industry that had been protected from imports since its establishment in the early 1950s. Prohibition of car importation was lifted but tariffs are still so high that imported cars do not constitute a threat to the position of the domestic automobile industry. Liberalization of much-demanded, imported electronic components and capital goods has not yet produced the effect of an increase in efficiency in production nor better quality cars. On the other hand, the same decision in relation to the importation of computers two years before the legal deadline resulted in a drastic reduction of R&D investment of some of the major computer producers in the country.

20 The refusal of taxpayers to collect their obligations was estimated to represent a loss somewhere in between a minimum of US$40 billion and US$160 billion per year: amounts equivalent to 10 to 40 per cent of Brazilian GDP. The administrative structure in charge of overseeing tax collection – itself plagued with corruption – was destroyed in the administrative reform and was not substituted by any more efficient structure. Scarce resources were made ever smaller and their destination decided without clear priorities. The government decided to bear the cost of the US$85.9 million spent repaying the foreign debt of northeastern sugar producers, rather than attempting to exact this from them. Moreover, while only 15 per cent of the US$750 million budget for the supplement of food for school children in 1992 had been effectively allocated up to August, US$1.5 billion was allocated for the construction of the Brasilia underground. Resources administered by the Bank of Brazil Foundation and destined to funding research, were diverted to serve the demands of traditional political elites in exchange for their support of the President and against his impeachment (Nussenzweig 1992).

REFERENCES

Abranches, S. H. H. (1978) 'The divided leviathan: state and economic policy formation in authoritarian Brazil', PhD dissertation, Ithaca, NY, Cornell University.

Adler, E. (1987) *The Power of Ideology. The Quest for Technological Autonomy in Argentina and Brazil*, Princeton, NJ, Princeton University Press.
Amadeo, E. (1978) 'Los Consejos Nacionales de Ciencia y Tecnología en America Latina: exitos y fracassos del primer decenio', in *Comercio Exterior*, vol. 28, no. 12, Mexico.
Barreto, A., Gusso, D. and Demo, P. (eds) (1990) 'Sistema educativo-cultural: uma visão prospectiva', *Para a Década de 90*, vol. 4, Brasilia, IPEA/IPLAN.
Bastos, M. I. (1991) 'Winning the battle to lose the war? The US/Brazilian dispute over the "informatics" Policy', University of Sussex, PhD dissertation.
Bastos, M. I. (1992) 'State policies and private interests: the struggle over information technology policy in Brazil', in Hubert Schmitz and José Cassiolato (eds) *Hi-tech for Industrial Development. Lessons from the Brazilian experience in electronics and automation*, London, Routledge.
Bastos, M. I. (1994) 'How international sanctions worked: domestic and foreign political constraints on the Brazilian Informatics Policy', *Journal of Development Studies*, vol. 30, no. 2, pp. 380–404.
Bell, M. and Cassiolato, J. (1993) *Technology Imports and the Dynamic Competitiveness of Brazilian Industry: The Need for New Approaches to Management and Policy*, Report for the Estudo da Competitividade da Industria Brasileira, University of Sussex, SPRU and CNPq and FECAMP/University of Campinas.
Bornstein, L. (1988) 'National autonomy and development strategy: informatics policy in Brazil', in Manuel Castells, Lisa Bornstein, Katharyne Mitchell, Rebecca Skinner and Jay Stowsky *The State and Technology Policy: A Comparative Analysis of the US Strategic Defence Initiative, Informatics Policy in Brazil, and Electronics Policy in China*, Berkeley, University of California, The Berkeley Roundtable on the International Economy (BRIE) Working Paper no. 37.
Botelho, A. J. (1989) 'Struggling to survive: the Brazilian Society for the Progress of Science (SBPC) and the authoritarian regime (1964–1980)', *Historia Scientiarum*, vol. 38, pp. 45–63.
Cardoso, F. H. (1977) 'Associated-dependent development: theoretical and practical implications', in Alfred Stepan (ed.) *Authoritarian Brazil*, New Haven, CT, Yale University Press.
Carvalho, R. Q. (1992) 'Why the market reserve is not enough: lessons from the diffusion of industrial automation technology in Brazilian process industries', in Hubert Schmitz and José Cassiolato (eds) *Hi-tech for industrial development. Lessons from the Brazilian experience in electronics and automation*, London, Routledge.
Cassiolato, J. E., Brunetti, J.L.A., and Paula, M.C.S. (1981) 'Desenvolvimento e perspectivas da política de ciência e tecnologia no Brasil', Brasilia, CNPq, mimeo.
Cassiolato, J. E., Hewitt, T. and Schmitz, H. (1992) 'Learning in industry and government: achievements, failures and lessons', in H. Schmitz and J. Cassiolato (eds) *Hi-tech for Industrial Development*, London and New York.
Castro, C. M. (1984) *É possível uma tecnologia 'Made in Brazil'?*, Brasilia, IPEA/CNRH.
Cavarozzi, M. (1992) 'Beyond transitions to democracy in Latin America', *Journal of Latin American Studies*, vol. 24, no. 3, pp. 665–84.
Conca, K. (1992) 'Technology, the military, and democracy in Brazil', *Journal of Interamerican Studies and World Affairs*, vol. 34, no. 1, pp. 141–77.

Congresso Nacional/CPMI (1992) *Relatório Final. Comissão Parlamentar Mista de Inquérito Causas e Dimensões do Atraso Tecnológico*, Brasilia.
Corbo, V. (1992) *Development Strategies and Policies in Latin America. A Historical perspective*, San Francisco, California, International Center for Economic Growth, Occasional Paper no. 22.
Coutinho, L. G., (ed.) (1991) *Desenvolvimento tecnológico da industrial e a constituição de um sistema nacional de inovacao no Brasil*, Campinas, Universidade de Campinas, Contract IPT/FECAMP.
Dahlman, C. J. and Frischtak, C. R. (1990) 'National systems supporting technical advance in industry: the Brazilian experience', Washington, World Bank, Industry Development Division, Industry and Energy Department, Industry Series Paper no. 32.
Deyo, F. C. (1987) 'Coalitions, institutions, and linkage sequencing – toward a strategic capacity model of East Asian development', in Frederic C. Deyo (ed.) *The Political Economy of East Asian Industrialism*, Ithaca, NY, Cornell University Press.
Diniz, E. and Boschi, R.R. (1989) 'Empresários e Constituinte: Continuidades e Rupturas no Modelo de Desenvolvimento Capitalista no Brasil', in Aspásia Camargo and Eli Diniz (eds) *Continuidade e Mudança no Brasil da Nova República*, São Paolo, Vértice Iuperj.
Dosi, G., Pavitt, K. and Soete, L. (1990) *The Economics of Technical Change and International Trade*, London, Harvester and Wheatsheaf.
Durham, E. R. and Gusso, D. A. (1991) 'Pós graduação no Brasil. Problemas e Perspectivas', paper prepared for the MEC/CAPES Seminário Internacional sobre Tendências da Pós Graduação, Brasília, 10–11 July.
Erber, F. S. (1979) 'Política científica e tecnológica no Brasil: uma revisão da literatura', in João Sayad (ed.) *Resenhas de Economia Brasileira*, São Paulo, Editora Saraiva.
O Estado de São Paulo (1990) 'Os caminhos da capacitação tecnológica', September 22.
Evans, P. B. (1989) 'Predatory, developmental, and other apparatuses: a comparative political economy perspective on the Third World State', *Sociological Forum*, vol. 4, no. 4, pp. 561–87.
Evans, P. B. (1992) 'The state as problem and solution: predation, embedded autonomy, and structural change', in Stephan Haggard and Robert Kaufman (eds) *The Politics of Adjustment. International Constraints, Distributive Conflicts, and the State*, Princeton, NJ, Princeton University Press.
Fajnzylber, F. (1990) *Unavoidable industrial restructuring in Latin America*, Durham and London, Duke University Press.
Fajnzylber, F. (1991) 'International insertion and institutional renewal', *CEPAL Review*, no. 44, pp. 137–66, Santiago, Chile.
Ferraz, J. C. (1992) *Modernização industrial à Brasileira*, Rio de Janeiro, IEI/UFRJ, Série Documentos no.7.
Ferraz, J. C., Rush, H. and Miles, I. (1992) *Development, Technology and Flexibility. Brazil Faces the Industrial Divide*, New York, Routledge.
Ferreira, J. P. (1980) 'Desenvolvimento científico e tecnológico – a experiência Brasileira', Rio de Janeiro, mimeo.
Fishlow, A. (1987) *Some Reflections on Comparative Latin American Economic*

Performance and Policy, Helsinki, World Institute for Development Economics Research of the United Nations University.

Fleury, A. C. (1989) 'The technological behaviour of state-owned enterprises in Brazil', in Jeffrey James (ed.) *The Technological Behaviour of Public Enterprises in Developing Countries*, London, Routledge.

Folha de São Paulo (1990) 'Programa tecnológico', September 14, page A-2.

Franko-Jones, P. (1992) *The Brazilian Defense Industry*, Boulder, CO, Westview Press.

Freeman, C. (1988) 'Japan: a new national system of innovation?', in G. Dosi, C. Freeman, R. Nelson, G. Silverberg and L. Soete (eds) *Technical Change and Economic Theory*, London and New York, Pinter Publishers.

Fritsch, W. (1990) 'A política industrial do novo governo: um passo a frente, dois para trás?', in Clovis de Faro (ed.) *Plano Collor. Avaliações e Perspectivas*, Rio de Janeiro, Livros Tecnicos e Científicos Editora Ltda.

Geddes, B. J. (1986) 'Economic development as a collective action problem: individual interests and innovation in Brazil', PhD dissertation, Berkeley, University of California.

Goldemberg, J. (1990) 'Os cientistas e o autoritarismo', *Folha de São Paulo*, July 19.

Granovetter, M. (1985) 'Economic action and social structure: the problem of embeddedness', *American Journal of Sociology*, vol. 91, no.3, pp. 481–510.

Guimarães, E. and Ford, E. (1975) 'Ciência e tecnologia nos planos de desenvolvimento', in *Pesquisa e planejamento econômico*, Rio de Janeiro, IPEA, vol. 5, no. 2, pp. 385–432.

Haggard, S. (1990) *Pathways from the Periphery. The Politics of Growth in the Newly Industrializing Countries*, Ithaca, NY, and London, Cornell University Press.

Hagopian, F. (1990) '"Democracy by undemocratic means?" Elites, political pacts, and regime transition in Brazil', *Comparative Political Studies*, vol. 23, no. 2, pp. 147–70.

Herrera, A. (1981) 'La planificación de la ciencia y la tecnología en America Latina: elementos para um nuevo marco de referencia', mimeo.

IEDI (1992) *A Nova Competitividade e Educação. Estratégias Empresariais*, São Paulo.

Johnson, C. (1981) *MITI and the Japanese Miracle*, Stanford CA, Stanford University Press.

Jornal de Brasília (1991a) 'Tecnologia só virá com garantias', March 7.

Jornal de Brasília (1991b) 'Publicação denuncia apartheid', July 7, p. 5.

Kaufman, R. R. (1990) 'Stabilization and adjustment in Argentina, Brazil and Mexico', in J. M. Nelson (ed.) *Economic Crisis and Policy Choice. The Politics of Adjustment in the Third World*, Princeton, NJ, Princeton University Press.

Kim, L. and Dahlman, C. J. (1992) 'Technology policy for industrialization: An integrative framework and Korea's experience', *Research Policy*, no. 21, pp. 437–52.

Klein, L. and Delgado, N. G. (1988) 'Recursos para a ciência: evolução e impasses', in *Ciência Hoje*, vol. 8, no. 48, pp. 28–33.

Kline, S. J. and Rosenberg, N. (1986) 'An overview of innovation', in Ralph Landau and Nathan Rosenberg (eds) *The Positive Sum Strategy. Harnessing Technology for Economic Growth*, Washington, National Academy Press.

Lin, C.-Y. (1989) *Latin America vs East Asia. A Comparative Development Perspective*, Armonk and London, East Gate Boork and M. E. Sharpe, Inc.

Lundvall, B.-Å. (1988) 'Innovation as an interactive process: from user–producer interaction to the national system of innovation', in G. Dosi *et al.* (eds) *Technical Change and Economic Theory*, London, Pinter.

Maciel, C. S. (1990) *Padrão de investimento industrial nos anos noventa e suas implicações para a política tecnológica*, IPT/FECAMP project report, Campinas, UNICAMP.

Martins, L. (1985) *Estado Capitalista e Burocracia no Brasil*, Rio de Janeiro, Paz e Terra.

Martins, L. (1986) 'The "Liberalization" of Authoritarian Rule in Brazil', in Guillermo O'Donnell, Philippe C. Schmitter and Laurence Whitehead (eds) *Transitions from Authoritarian Rule*, Baltimore and London, Johns Hopkins University Press.

Martins, L. (1990) 'A autonomia política do governo Collor', in Clovis de Faro (ed.) *Plano Collor. Avaliações e Perspectivas*, Rio de Janeiro, Livros Técnicos e Científicos Editora Ltda.

Medeiros, J. A. (1991) 'Novas parcerias no setor espacial fortalecem a empresa', *Gazeta Mercantil*, April 30.

Medeiros, J. A. (1990) 'As duas faces dos polos tecnológicos', *Folha de São Paulo*, February 26.

Meyer-Stamer, J. (1993) 'Comprehensive modernization on the shop-floor: a case study on the machinery industry in Brazil', in Klaus Esser, Wolfgang Hillebrand, Dirk Messner and Jörg Meyer-Stamer, *International Competitiveness in Latin America and East Asia*, Berlin, Frank Cass.

Ministério da Ciência e Tecnologia – MCT (1988) *Síntese histórica organizacional*, Brasília.

Ministério da Economia, Fazenda e Planejamento do Brasil (1990) *Diretrizes gerais para a política industrial e de comercio exterior*, Brasília.

Ministério da Economia, Fazenda e Planejamento do Brasil (1990) *Programa Brasileiro da qualidade e produtividade*, Brasília.

Mowery, D. C. (1992) 'The US national innovation system: Origins and prospects for change', *Research Policy*, no. 21, pp. 125–44.

Mowery, D. C. (1994) *Science and Technology Policy in Interdependent Economies*, Boston, Dordrecht, London, Kluwer Academic Publishers.

Mundim, M. and Penteado, N. (1990) 'Governo vai isentar de IR quem investir em tecnologia', *Jornal de Brasília*, September 2.

Musa, E. V. (1991) 'Uma velha e boa parceria', *O Estado de São Paulo*, March 5.

Nakatani, A. (1991) 'A tecnologia necessária', *O Estado de São Paulo*, February 6.

Nelson, J. M. (ed.) (1990) *Economic Crisis and Policy Choice. The Politics of Adjustment in the Third World*, Princeton, NJ, Princeton University Press.

Nunes, E. O. (1984) *Bureaucratic Insulation and Clientelism in Contemporary Brazil: Uneven State-Building and the Taming of Modernity*, PhD Dissertation, Department of Political Science, University of California, Berkeley.

Nunes Amorim, C. L. (1991) 'A América Latina e as diferentes formas de cooperação técnico-científica', in Secretaria de Ciência, Tecnologia e Desenvolvimento Econômico de São Paulo, and CNPq, *Ciência e Tecnologia na América Latina*, São Paulo.

Nussenzweig, H. M. (1992a) 'Governo federal liquida a ciência', *Folha de São Paulo*, April 13.

Nussenzwig, H. M. (1992b) 'Marcílio, o exterminador do futuro', *Folha de São Paulo*, August 21.

Onis, Z. (1991) 'The logic of the developmental state', *Comparative Politics*, vol. 24, no. 1, pp. 109–26.

Pack, H. and Westphal, L. E. (1986) 'Industrial strategy and technological change', *Journal of Development Economics*, no. 22, pp. 87–128.

Proença, D. (1987) *Tecnologia militar e os militares na tecnologia: o caso da 'política nacional de informatica'*, MA thesis, COPPE/OFRJ, Rio de Janeiro.

Proença, D. (1990) 'Guns and butter? Arms industry, technology and democracy in Brazil', *Bulletin of Peace Proposals*, vol. 21, no. 1, pp. 49–57.

Rosa, L. (1985) *A política nuclear e o caminho das armas atômicas*, Rio de Janeiro, Zahar Editores.

Rowe, James W. (1969) 'Science and politics in Brazil: background of the 1967 debate on nuclear energy policy', in Kalman H. Silvert (ed.) *The Social Reality of Scientific Myth*, NY, American Universities Field Staff.

Rueschemeyer, D. and Evans, P. B. (1985) 'The state and economic transformation: toward an analysis of the conditions underlying effective intervention', in Peter B. Evans, Dietriech Rueschemeyer and Theda Skocpol (eds) *Bringing the State Back In*, Cambridge, Cambridge University Press.

Salm, C. (1992) *Dois textos sobre educação e transformações tecnológicas*, Rio de Janeiro, IEI/UFRJ, Textos para Discussão no. 280.

Schmitz, H. and Cassiolato, J. 'Fostering hi-tech indusries in developing countries: introduction', in H. Schmitz and J. Cassiolato (eds) *Hi-Tech for Industrial Development*, London and New York, Routledge.

Schneider, B.R. (1991) 'The career connection: a comparative analysis of bureaucratic preferences and insulation', Department of Politics, Princeton University, mimeo.

Schneider, B. R. (1991a) 'Brazil under Collor: anatomy of a crisis', *World Policy Journal*, vol. 8, no. 2, pp. 321–47.

SCT/CNPq (1990) *Guia de fontes de financiamento à ciência e tecnologia*, Brasilia.

Secretaria da Ciência e Tecnologia da Presidência da República (1991) *A política Brasileira de ciência e tecnologia 1990/95*, Brasilia, 2nd edn.

Shapiro, H. (1994) *Engines of Growth. The State and Transnational Auto Companies in Brazil*, Cambridge, Cambridge University Press.

Schwartzman, S. (1979) *Formação da Comunidade Científica no Brasil*, Rio de Janeiro, FINEP and Companhia Editora Nacional.

Schwartzman, S. (1985) *High Technology vs. Self-Reliance: Brazil Enters the Computer Age*, Rio de Janeiro, IUPERJ.

Schwartzman, S. (1989) 'Ciência e tecnologia na nova republica', *Ciência e Cultura*, vol. 9, no. 50, pp. 62–9.

Souza Paula, M. C. (1991) 'Oportunidades e entraves ao desenvolvimento tecnológico no Brasil: a experiência da industria aeronáutica e indústria farmacêutica', PhD dissertation, Universidade de São Paulo.

Stepan, A. (1985) 'State power and the strength of civil society in the southern cone of Latin America', in P. B. Evans, D. Rueschemeyers and T. Skocpol (eds) *Bringing the State Back In*, Cambridge, Cambridge University Press.

Suzigan, W. (1991) *A indústria Brasileira após uma década de estagnação*, paper presented in the Sextas Jornadas Anuales de Economia, Banco Central del Uruguay, Montevideo, 4–6 November.

Vaitsos, C. (1990) *The Needs and Possibilities for Cooperation Between Selected Advanced Developing Countries and the Community in the Field of Science and Technology. Country Report on Brazil*, prepared for the Strategic Analysis in Science and Technology Unit (SAST) of the Directorate-General for Science, Research and Development of the Commission of the European Communities.

Vargas, J. I. *et al.* (1986) *Avaliação do programa nuclear Brasileiro*, Brasília, published by Academia Brasileira de Ciências, Rio de Janeiro, 1990.

Villaschi, A. (1992) 'Is Brazil ready to leap-frog under the IT techno-economic paradigm? Some questions regarding its education system', paper prepared for the Second International Conference on Science and Technology in Third World Development, University of Strathclyde, Glasgow, 5–7 April.

Vogt, C. (1990) 'Universidades e programa tecnológico', Jornal do Brasil, October 8.

Vogt, C. (1991) 'Crise e qualificação tecnológica', *Jornal do Brasil*, May 28, no. 1, p. 11.

Wade, R. (1990) *Governing the Market. Economic Theory and the Role of Government in East Asian Industrialization*, Princeton, NJ, Princeton University Press.

Willis, E. (1986) *The State as a Banker: The Expansion of the Public Sector in Brazil*, PhD dissertation, University of Texas at Austin.

4 Harnessing the politics of science and technology policy in Mexico

Alejandro Nadal Egea[1]

1 INTRODUCTION

In which manner science and technology (S&T) policy initiatives are triggered and carried out is a question of critical importance. The study of how different actors play their roles in launching and implementing diverse policy initiatives, and the circumstances surrounding their actions (including the institutional framework) provides important insights for the design of viable science and technology policies. The viability and effectiveness of a particular S&T policy are heavily dependent upon the relative composition of political forces and coalitions encompassing these agents' actions. In addition, the nature of the political regime and its proneness to manipulation by interest groups, its flexibility or its rigidities, its capabilities for response to different demands from political actors and of international context – all of these are determinant variables of the applicability of S&T policy.[2]

This chapter is not concerned with the debate on whether science can change politics or if politics affects scientific endeavour. It is clear that political conditions cannot affect the internal dynamics of scientific research, its methods and its results. And it is equally clear that science cannot in itself determine the nature of a political regime. Employing technical jargon in political discourse cannot transform political decision-making into a more rational activity. But political forces can and frequently do decide which research areas are to be investigated and which are to be relegated as secondary items in the research agenda. Political forces can also condition the manner in which decisions concerning science and technology are adopted and implemented. Decisions governing the allocation of resources for the acquisition of technologies, whether through domestic R&D or licensing agreements, emanate from a political process. And the choice of the array of specific policy instruments through which these decisions are actually implemented is dependent on the political forces acting in this environment. The analysis of some of these processes is the object of this chapter.

In recent years, the technology and industrial policy debate in Mexico has systematically ignored the fact that export-promotion strategies in East Asian countries relied heavily on protectionism and state intervention. As a result, one of the main issues being overlooked is that in several economic success stories, such as the Republic of Korea, and, of course, Japan, export-promotion strategies depended critically on stiff government controls and an intensely regulated economic environment. A highly protected agricultural production coexisted with strict controls over direct foreign investment and transfer of technology operations and, in the case of Japan, market structures acted as extremely effective non-tariff (and non-quantitative) trade barriers. Of course, a more balanced income distribution was an important contextual factor of these strategies. In these success stories, targeted industries received generous direct and indirect support through preferential access to capital and subsidies. A prerequisite for this in Japan, Korea and Taiwan was the existence of a state apparatus strong enough to establish patterns of economic co-ordination and control in a rigid manner for sufficiently long periods of time. Not a single social group was strong enough to act as countervailing power, challenging the structure of these political organizations. Each respective state could assume a clear and undisputed leadership in the economic field.

Mexico is also a country with a strong state. But this is as far as the comparison goes because the state was unable to establish and implement a long-term, coherent industrial cum technology policy required for a more successful outcome in the international economic arena. Why is it that in the case of two strong states, say Korea and Mexico, even when both share some important authoritarian features, we are faced with two entirely different experiences as far as industrial and technology policies are concerned?

The explanation may lie in the structural differences between authoritarian states. Not all authoritarian states have the same goals and *modus operandi*. Since the 1920s, the state goals in Mexico have been dominated by the priority of keeping political stability. This was the leitmotif of the creation in the mid-1920s of the state party, now the Institutional Revolutionary Party (PRI). The PRI is the state party in that it has direct and unlimited access to public resources for its campaigns, the media are controlled by its government officials, and it organizes and controls the electoral processes at the municipal, local and federal levels of government. Originally conceived as the political instance where the various factions that had taken part in the Mexican Revolution could negotiate and reach acceptable political agreements, the PRI has become a formidable political machinery with its sole purpose aimed at maintaining its officials in office.

This inclination to place political stability as the dominant priority may be the heritage of a very complex national structure, comprised of strong cultural and ethnic differences, and a large territory where the relative weight of regional power blocs has become an important variable. Having this key priority dominate all other considerations triggered the creation of a corporatist structure in which peasants and workers were brought into the body politic through centralized organizations which were appendages of the state party, the PRI. Peasants were grouped together in a well-organized confederation of peasant organizations, the CNC; yet the main objective of the agricultural system was not to produce more along with higher productivity (although these goals were not actually discouraged), but to keep peasants under firm political control. Unionized workers, on the other hand, were controlled through a centralized confederation of unions, the CTM. Both the CNC and the CTM were the corporatist branches of the PRI and, together with a loosely held alliance of urban associations, they were the key to the PRI's predominance. Finally, the political expression of Mexican nationalism has not been channelled towards long-term *economic* objectives. The stridency of intense nationalistic moments in recent history (expropriation of the foreign oil companies in 1938 and of the utilities in 1960) has been more related to certain key decisions on securing 'national control' of resources considered strategic assets. The political establishment has used these moments to gain public support from workers, peasants and the urban middle class, to further the cause of political stability.

However, since the 1980s the PRI has become the scene of renewed tensions between the old school of politicians and the new group of technocrats (most of which are economists and political scientists). The result has been that given its structure as a political conglomerate, the state party has become more subordinated than ever before to the decisions of the executive power. This evolution may have been due to a variety of reasons. During the first four or five decades of this century the economy became more diversified, creating very different power sources and varied constituencies. Throughout this period economic inequality has remained a pervasive force in Mexico's development. In summary, the dominant goal of political stability has been transformed into a short-term objective of maintaining the state party officials in power by all available means. This has had a negative impact on the realm of technology and industrial policies.

This chapter is structured as follows. Section 2 describes the main features of the scientific and technological system (including the S&T policy component) in Mexico. Section 3 focuses on the relation between the scientific community and the federal government; the analysis in this section centres on the political constraints affecting the Science and Technology Council (CONACYT) since its creation and the efforts of the federal government to enrol the political support of the scientific community. Section 4 considers a different aspect of technology policy in Mexico: the interaction between private interest groups and government officials in the realm of protectionism, transfer of technology and industrial property. Section 5 offers concluding remarks on the future of S&T policy in Mexico, considering that an open economy cannot survive *without* a well-designed one.

2 THE SCIENTIFIC AND TECHNOLOGICAL SYSTEM IN MEXICO (1950–92)

The main characteristics of the science and technology (S&T) system in Mexico can be summarized as follows. First, total R&D expenditures are very low: US$870 million dollars in 1993, representing only 0.48 per cent of GDP. In addition, public sector participation in R&D expenditures is dominant. According to official data, the public sector's contribution to total R&D expenditures was 85 per cent in 1984. There is even evidence suggesting this figure underestimates the importance of the public sector. By any standards, this is a powerful testimony to weak private participation in total R&D spending. In 1991–2, the Science and Technology Council (CONACYT) released data from a survey indicating that the participation of the public sector had decreased to 78 per cent, while the private sector's share had increased to 22 per cent (up from only 10 per cent in the late 1970s). This is equivalent to almost US$ 232 million and although it still is a very skewed representation of the sources contributing to R&D spending, it would come closer to other national experiences.[3] However, the official data for 1991 have been ardently contested by the academic community and there are reasons to believe that the increment of the private sector's share in total R&D spending has been grossly overestimated. For one thing, the methodology used and the sample of firms covered by CONACYT's survey have not been disclosed. In addition, in the recent past CONACYT has shown a tendency to overstate the impact of an open economy on privately funded R&D and this has partially eroded its credibility.

Second, R&D is heavily concentrated in a few institutions. The academic sector is responsible for approximately 45 per cent of federal R&D expenditures: most of these resources go to a handful of higher education and research institutions, of which the lion's share is received by the National Autonomous University (UNAM) and the Polytechnic

Table 4.1 Federal R&D expenditure in Mexico, 1970–91 (millions 1980 pesos)

(1) *Year*	*(2)* *R&D expenditures*	*(3)* *ROG (%)*
1970	3,332	–
1971	7,147	114.5
1972	8,557	19.7
1973	9,728	13.7
1974	10,124	4.1
1975	9,334	–7.8
1976	9,477	1.5
1977	9,806	3.5
1978	12,524	27.7
1979	14,135	12.9
1980	19,193	35.8
1981	22,264	16.0
1982	20,245	–9.1
1983	14,674	–27.5
1984	17,645	20.2
1985	17,431	–1.2
1986	16,615	–4.6
1987	13,478	–18.8
1988	13,158	–2.3
1989	13,954	6.0
1990	15,755	13.0
1991	20,153	28.0
1992 (p)	20,591	2.1

Notes: Column 2 indicates R&D effective expenditures incurred by public sector agencies, including federally funded R&D executed by private firms.

Sources: 1970–80, Academia de la Investigación Científica; 1980–92, Consejo Nacional de Ciencia y Tecnologia (1992b).

Institute (IPN). A few large public sector institutions in energy, agriculture, health and other sectors, receive another large component of total R&D expenditures. A corollary of the above is the extreme regional concentration of research institutions.

Third, R&D expenditures at the aggregate level in Mexico in the period 1970–90 reveal violent fluctuations (Table 4.1). There was an increasing trend between 1970 and 1981: the per annum growth rate for this period was 18.8 per cent which in itself is an extraordinary performance given that the economic growth rate for this period was 4.5 per cent. The growth rate (for R&D expenditure) started to slow down drastically in 1981, and there was a dramatic downward trend after 1982, followed by a slow recovery in the late 1980s. By 1991 official R&D expenditures reached the levels of the 1982 fiscal year. The 1982 crisis revealed how fragile the government's commitment to R&D was, as all federal agencies implemented severe cuts in the R&D budget.

Fourth, S&T policy has not been stable. In general, the allocation of R&D resources to different objectives, sectors and institutions has not been dictated by a conscious pattern of well-defined priorities and long-term objectives. It has often been determined by political leverage and access to channels of power. R&D resource allocation on a stable basis for certain sectors has also been an outgrowth of political needs, as in the case of the health, agricultural, energy and public works (transportation, communications, hydroelectric and irrigation infrastructure, etc.) sectors. In these cases the state has faced immediate fundamental requirements and had to react accordingly.

Interestingly, these cases are good examples of areas in which R&D activities have benefited from its linkage with the supply of goods and services, leading to quite positive results in technology development.

The medical profession has been quite active in organizing a set of solid research capabilities. In this task, it found a powerful ally in the pharmaceutical industry, which as a well-organized lobbying body has never lost sight of the importance of the purchasing power of the Mexican social security system. Originally, state political obligations in terms of public health were the basis for the pattern of resource allocation; but during the 1980s the macroeconomic imperative of reducing public spending carried even more weight. In the 1950s and 1960s, the share of the health sector in total government R&D spending was approximately 20 per cent; but later this share stabilized at 10 per cent and during the 1980s, the health sector saw its share further reduced to 6.5 per cent. Today, not more than 4.5 per cent of total federal R&D disbursements are allocated to health.

Agricultural research also received support because the state had crucial political obligations with the peasant population stemming from the 1910–20 Revolution. However, in the end, the bulk of agricultural research went to commercial production and not to peasant agriculture. The role of private international organizations and large commercial enterprises dominating seed production and marketing, as well as the production of chemical inputs, was essential to this process.[4] As with the health sector, the share of agriculture in total federal R&D expenditures oscillated around the 25 per cent level; in the 1970s this was reduced to approximately 20 per cent and it has reached a new all-time low at 10 per cent. Evidently, the community of agricultural researchers does not have the power to mobilize enough political support to obtain higher allocations. During 1991–92 important legal changes were introduced in the area of land ownership. The rationale was to introduce precise definitions of what constituted private property rights in view of effectively terminating the *ejido* (common land) system. Thus, the low R&D allocations could also reflect the policy priority of leaving agricultural R&D to market forces.

In the energy sector the state has had to act in all energy-related fields: oil, electricity and nuclear energy. It created the Mexican Petroleum Institute (IMP) to function as an engineering firm and specialized laboratory for the state-owned oil monopoly PEMEX. It also created two special research institutes devoted entirely to meet the research requirements of the Federal Electricity Commission (CFE) and the Atomic Energy Commission (AEC). In the case of PEMEX, the IMP has helped attain a high degree of technological sophistication in extraction and refining technologies, although there is still much to be done in the realm of off-shore exploration and drilling. The political commitments of the state ensuing from the 1938 expropriation help explain this unique example of a durable allocation of R&D resources in a specific sector. This priority did not stem from a self-conscious act of science and technology planning, but from a political mandate to consolidate independent resource management practices. To what extent this goal has been attained is a different matter, as this depends on powerful downstream market forces. Today, this mandate is being modified as even PEMEX is entering the list of privatized public firms, so the research mission of IMP remains to be confirmed.

In the early years of the Revolution, the need for public works in irrigation and transportation became apparent. Consequently, federal administrations launched a massive long-term programme of public investment in transportation (mainly through the road network) and in large hydroelectric dams. Some of these became showcases of the ability of private civil engineering firms to launch and execute large and complex projects involving sophisticated construction technologies and machinery. However, the provision of core capital goods (large-scale turbines, electromechanical controls, pumps and valves, etc.) remained an area controlled by foreign contractors. The large civil engineering firms in Mexico joined efforts to create the Engineering Institute of the National Autonomous University (UNAM) which has served, since the mid-1950s, as their research laboratory. The political dimension of the large infrastructure projects is responsible for the relatively stable pattern of R&D resource allocation in this case, although R&D was carried out for private firms in the laboratories they donated to public universities. The prospect of obtaining public contracts by these firms was an important incentive for establishing this peculiar R&D organization.

Beyond these cases there has not been a well-defined general *strategy* for the allocation of R&D resources; this has led to a dispersion of resources involving many negative implications. This is particularly clear with respect to the manufacturing industry as a whole: no decisive effort for determining long-term priorities has been carried out. There has not been an institution capable of defining priorities, co-ordinating private firms in the different dimensions of long-term implementation of strategically important industrial projects, including financing, access (assimilation) of technology, construction, production, marketing, etc. The government was unable to orient private in-house R&D in definite directions; it was only able to carry out some promotion and regulatory activities in relation to privately financed R&D without any clear definition of sectoral priorities. Allocation of R&D resources in the health, agricultural, energy and infrastructure sectors was carried out as a consequence of state political commitments. As regards infrastructure, R&D was conducted in university laboratories for firms expected to preserve their competitive advantage in bidding for public contracts.

Fifth, the dominant paradigm in S&T policy has been to provide resources for R&D in all branches of activity, with the hope that some sort of 'trickle-down' effect will ensue. There are important efforts to link R&D in academic centres with production activities, and the main academic institutions have set up offices to bridge the gap between R&D and productive enterprises. However, the main axes of the government's science policy have been related to strengthening infrastructure, training of human resources and, in general, promoting R&D activities in all sectors.

To the extent that Mexico has an important number of research laboratories and institutes, and a group of co-ordinating bodies, of which the Science and Technology Council is the most important, it may be tempting to conclude that Mexico has a scientific and technological system. However, the systemic nature of this array of organizations is rather loose and there are no internal relations determining the general evolution of the system.

Sixth, the instruments used for the implementation of S&T policies have widely varied in nature, ranging from credits and financial assistance on the promotional side, to controls of licensing agreements on the regulatory side. The actual operations of these instruments relied on high discretionary powers of officials in charge and this has made them prone to manipulation by interest groups. Non-selectivity and discretionary powers have dominated the style of S&T policy instruments in the past. Reliance on the market mechanism will most probably make non-selectivity a durable feature of these policy instruments in the future.

In order to complete this general frame of reference of the science and technology system in Mexico, the lack of a democratic government and a juridical institutional framework must be emphasized. This has made policy design and implementation extremely vulnerable to manipulation by interest groups. Several conclusions emanate from this fact. One is that interest groups do not act through the normal channels of political parties in Parliament or Congress. Thus, a sizeable amount of literature on social action and rational choice is simply irrelevant in the Mexican context. Manipulation is thus far removed from the public eye. Another conclusion is that there is almost no accountability for the actions of public officials (nor for the actions of interest groups either). The legislative branch of government is entirely controlled by the executive power and does not conduct impartial open hearings; all hearings are rigged with the participation of benign critics of the policies being considered. This explains how the recent enactment of the Industrial Property Act was carried out with practically no debate. This also explains how the regulations on foreign investment can contradict parts of the law on foreign investment and how this abnormal situation can last for several years without public outcry.

Civil society is not empowered to start judicial actions against public authorities for violations of the law.[5] Matters such as the 'right-to-know act' are almost totally ignored in Mexico today. This brings us to the last and perhaps most important point in this section: a weak rule of law, a lack both of accountability and of a democratic political system all undermine the possibility of designing and implementing a sound and robust S&T policy which is not prone to manipulation. Science policy can be present in strong authoritarian states, and in fact it can be argued that Mexico, just as Korea is an authoritarian state (although of an entirely different foundational structure). But this does not mean that authoritarianism is the *sine qua non* condition for a successful S&T policy, or even an advantage for it. If there is no rule of law, no accountability and little legitimacy stemming from the electoral system, authoritarianism may be synonymous with a system where only large corporations and powerful interest groups have almost exclusive access to the sphere of political influences. Absence of rule of law and a weak division of powers are accompanied by proneness to manipulation. Lack of accountability brings about want of the continuity required by large science and technology projects. To a considerable degree, these elements are already present (and will probably continue to be present) in Mexico as a result of the specific nature of its political system.

3 NOBODY'S CONSTITUENCY: THE SCIENTIFIC COMMUNITY. CONACYT AND SCIENCE AND TECHNOLOGY POLICY IN MEXICO

In 1968, Mexican society was shocked by the most serious political movement since the years of the Revolution. The movement was spearheaded by thousands of university students, but the dominating issues went beyond the realm of a student conflict and included more general demands for greater freedom and democratization of the Mexican state. A mixture of pseudo-negotiations and repression was the answer of the political establishment, culminating on 2 October 1968 in the massacre (executed by army and police units) of dozens of participants in a peaceful demonstration in Mexico City, effectively terminating the movement ten days before the inauguration of the Olympic Games.

The 1968 movement was over, but it sent shock waves through the political establishment, clearly showing that the time for some kind of political reform had come. Alongside this political reform came the recognition by the political establishment that the academic community (and this included all scientists and engineers) had to have channels of access to political leverage and financial resources. The National Autonomous University (UNAM) and the National Polytechnic Institute (IPN) had been the two central foci of the movement, suffering the consequences as the army occupied the entire campus of UNAM and aggressively stormed different units of IPN. At one point, the Rector of UNAM, a well-known civil engineer, Javier Barros Sierra, headed a peaceful demonstration protesting the army's violation of the university's autonomy. The university and polytechnic had somewhat been transformed into a battlefield and scientists, engineers and support personnel in academic laboratories had been too close to the main actors in the student movement to remain on neutral ground. Most of the scientific and technological establishment was shocked and took a clear position behind the student movement.

With the start of the process for presidential succession in 1969-70, the National Institute for Scientific Research (INIC), a body created in 1950 and dependent on the Ministry of Education, was charged with the task of carrying out a series of studies in order to define the main lines of a national policy for science and technology.[6] The Rector of the National University and the Director of the National Polytechnic Institute had already played a prominent role in a crucial conference on science, technology and productivity held in 1967. The central resolution of this conference was a request to create 'a committee for the study and promotion of science and technology', integrated by the Rector of UNAM, the Director of IPN and the Secretary-General of INIC. The committee's work centred on the preparation of a draft law designed to reorganize INIC but its work was interrupted by the events of 1968.

According to the official record of CONACYT's foundation, the scientific community was highly concerned about the lack of linkages between scientific and technological research and productive activities. However, the document does not mention the fact that after the fateful year of 1968, professional politicians were highly concerned about the lack of a firm grasp or political control over scientists and researchers. It is also surprising that sociologists and political analysts have been slow to recognize this; only marginal attention has been given to the fact that political considerations were important for the creation of CONACYT during the first year of President Echeverría's mandate (see for example, Casas and Ponce 1986) and no attention has been accorded to the key role of the academic community in the 1968 movement.

It has become common in Latin American research to mention frequently that the recommendations of UNESCO during the 1960s to create bodies for the formulation and implementation of a comprehensive science policy was the direct precedent for the establishment of science councils in Latin America, and of CONACYT in Mexico. There is no doubt that these recommendations stemming from an international organization such as UNESCO were important; maybe even more so when viewed in terms of injecting a sense of urgency in the political establishment to maintain control of such a potentially important group as the scientific community. This was a social group that to a large extent had become nobody's constituency and, clearly, by the late 1960s, it had become of strategic importance to have this community well under some sort of control.

Two interesting corollaries of the above are the following. First, as the Mexican political establishment could only impose *manu militari*, its solution to the 1968 conflict sent a clear signal that relations with the scientific and technological communities could not be carried out on a 'business as usual' basis. In fact, the message issued from the repressive response was not one of strength, as a shortsighted view would have it, but of political weakness. Second, the quest for control and gaining legitimacy over this constituency continues to this date. Of course, the channels of political action have changed, but 1970 was a watershed date: since then, the Mexican political establishment, both in quarters of the official party and of the opposition, has shown concern for the orientation of the scientific community and its political leanings.

During the first quarter of 1969, the Ministry of the Presidency convened a series of meetings with the directors and top staff of the main research institutions in Mexico. These meetings were part of a general consultative effort to gather information on the scientific community's viewpoint regarding the requirements for a comprehensive science policy in Mexico. The outcome of this process was an executive order to INIC, charging this institute with the task of carrying out the necessary steps to establish the institutional base for the development and implementation of S&T policy in Mexico. INIC launched an ambitious programme of meetings and studies aimed at answering three critical questions:

1. What are the conditions under which scientific and technological research in Mexico takes place?
2. How can the main obstacles impinging on S&T research be surmounted?
3. Which are the instruments that will enhance and strengthen S&T research in Mexico?

By the end of 1970, INIC produced a final report with a series of recommendations. Occupying a central position was the consideration that serious problems were slowing down scientific development, and foremost among these was the 'lack of understanding and support for scientific research'. This consideration lends support to the main hypothesis here: the political establishment was willing to provide support and understanding, but it would expect something in return. INIC's resolutions would incorporate this payoff in terms of the 'orientation' of the R&D effort. The report concluded that the lack of linkages between R&D and productive activities was a major problem that had to be solved; it also incorporated new ideas related to linkages between the R&D effort and 'national development goals' such as health, nutrition, housing, education, natural resources and economic development. As will be seen later, the attempt to orient R&D towards certain 'national development goals' was shortlived. But, several years afterwards the state finally obtained stricter levels of political control over the scientific community through the reduction of real income accruing to researchers.

The end result of this process was the creation in 1971 of the National Council for Science and Technology (CONACYT). Since that year, CONACYT has been the focus of dialogue and communication between government and the scientific community. Of course, other important channels of communication have existed (such as the Academy for

Scientific Research, AIC), but CONACYT is the main centre for negotiations regarding financial resources and general orientation of R&D in Mexico.

Formally, CONACYT was set up (in accordance with recommendations of the original INIC study) as a decentralized body responsible for the design and implementation of S&T policy in Mexico. But the Council's activities had two dimensions: first, as the obligatory advisor to public sector agencies (from ministries to state-controlled firms) on all matters regarding science and technology, and, second, as the executor of so-called 'auxiliary' activities. These auxiliary activities were defined as the additional (financial and logistic) support CONACYT would be enabled to provide to scientific and technological institutions.

Under this provision of its enactment law, CONACYT was to help influence the orientation of decisions regarding R&D, technology transfer and manpower training. Interestingly enough, purchases of capital goods embodying technology, as well as basic and detail engineering, are not covered by CONACYT's law. Although the law did not specify the set of criteria that was to be used in this advisory capacity, three main lines of action were actively envisaged by CONACYT's early administrations. The first was to influence decision-making on R&D in order to bring about a greater affinity between lines of research and 'national development problems'. The second was to try to establish better linkages between the public sector and the domestic scientific research effort. This was related to the question of adapting technology to local conditions. The question of direct costs of technology transfer operations was considered a third possible dimension for CONACYT's intervention.

In its capacity as obligatory adviser to the public sector, CONACYT failed to influence the substance of S&T decisions in Mexico. The reason is quite simple and could not have escaped the drafters of CONACYT's organic law: the Council was a decentralized body and not a ministry. In the not too complicated universe of Mexico's public sector, where decision-making is heavily centralized, the relative weight of a decentralized body is almost negligible when confronted with monster state-owned firms or state-controlled entities such as Petroleos Mexicanos (PEMEX), Comision Federal de Electricidad (CFE), or the Instituto Mexicano del Seguro Social (IMSS). And *vis-à-vis* giant ministries such as the Ministry of Energy and Public Works, the Ministry of Agriculture, or the Ministry of Education, CONACYT was (and still is) helplessly dwarfed and could not possibly influence their science and technology decisions.

Of course, being a ministry is a necessary, but not a sufficient, condition for political clout. And it remains to be seen if a Ministry of Science and

Technology would have the power needed to effectively influence S&T decisions in the public sector. The point at hand is that even to a casual observer of Mexico's public sector, CONACYT's function as an obligatory adviser could not have been considered a realistic one. Thus, at best, establishing CONACYT with this advisory power was a harmless deed. Perhaps a more accurate interpretation is that this was designed to entice the scientific community into providing the needed political support for CONACYT's infant years, in exchange for an illusion of real power to influence the S&T decision-making process of Mexico's public sector.

The second avenue for CONACYT's activity was presented as the way to influence trends in R&D through the allocation of additional resources to the scientific and technological effort. However, the percentage share of resources allocated to R&D by CONACYT are not enough to achieve this result: as Table 4.2 reveals, this percentage does not exceed 10 per cent during the entire decade. Most of the federally funded R&D is carried out in the institutions of the academic sector (20 per cent of federal R&D spending goes to three institutions in Mexico City), in energy and agriculture. This participation in total R&D expenditures is consistent with the trend during the first years of CONACYT's existence when the share was approximately 11 per cent (Nadal 1977: 24). The question of CONACYT's relative weight in total R&D expenditures ceased to be a critical issue in the latter half of the 1970s because, as we shall see, S&T policy changed course in a fundamental way. After 1977, no S&T policy priorities were defined and, thus, CONACYT's strategy no longer influenced the direction of R&D through the allocation of supplementary resources.

CONACYT's power to influence the allocation of resources towards certain research priorities was always seen as a threat to the autonomy of the public universities.[7] From a political perspective, the scientific community looked upon CONACYT during the 1970s both as a potential straitjacket and as a rich instrument to obtain extra resources.

It is not surprising to observe that a struggle followed between government officials and the hierarchy of scientific institutions to resolve this contradiction. This political struggle has taken different forms during CONACYT's lifetime. Two extreme forms are the planning experience of 1975–6 and the current view which effectively surrenders all decision-making of priorities to market forces. A brief analysis of these two processes is helpful to understand the political dynamics determining S&T policy in Mexico.

During 1975–6 an ambitious planning exercise was undertaken by CONACYT to define a framework for a comprehensive science and technology plan. The plan, projected as an indicative framework, was to serve as a

Table 4.2 CONACYT'S share of federal expenditure in science and technology, 1980–92 (millions 1980 pesos)

Year	(1) Federal S&T budget (a)	(2) CONACYT's S&T budget (b)	(3) CONACYT's share (2)/(1)
1980	19,193	1,563	8.1
1981	22,264	2,020	9.0
1982	20,245	1,977	9.7
1983	14,674	1,609	10.9
1984	17,645	1,600	9.0
1985	17,431	1,711	9.8
1986	16,615	1,242	7.4
1987	13,478	1,071	7.9
1988	13,158	1,104	8.3
1989	13,954	1,040	7.4
1990	15,755	1,407	8.9
1991	20,153	2,044	10.1
1992 (p)	20,591	4,229	20.5

Notes:
(a) Federally supported R&D in government, public sector laboratories, universities and private enterprise
(b) R&D budget administered by CONACYT. This column does not include administrative and planning expenditures of CONACYT
(p) Preliminary figures

Sources: Tables I.1 and IV.5, Indicadores Actividades Cientificas y Tecnologicas, CONACYT, Mexico, 1992

general reference for S&T policy. Early in the process the participatory nature of the plan was defined as a critical element; consequently, researchers from all major centres and institutions, both academic (i.e. autonomous) and from the public sector, were invited to take part in special committees. The plan concentrated on the applied research and technological side of the spectrum of R&D activities. The planning exercise was organized in a complex, multi-tiered format, with strategic guidelines for the science, technology and human resources dimensions of the system, as well as sectoral priorities for 'technological development' in the following fields: food technologies; agriculture, livestock and forestry; fisheries; manufacturing industry; mining; energy; transportation and telecommunications; urban development and housing; medicine and health; education.

The plan also had a special chapter on scientific development covering two blocks of disciplines: exact and natural sciences, and the social sciences. In the rhetoric of the final document, 'scientific development' meant basic research. CONACYT's officials knew they were touching the highly sensitive issue of the 'academic autonomy' question because this kind of research is intensely concentrated in the National University and the Polytechnic Institute. The priorities for these activities were defined in very general terms (CONACYT 1976: 123–4) including: establishing linkages between basic research and the rest of the R&D effort in the country, basic research emanating from research on 'national problems of development', as well as supporting basic research in generic technologies that had a higher potential for application. But the first two priorities are important from the political perspective: promotion of research in areas where Mexican researchers have already acquired international recognition, and support of high-quality scientific research (carried out by researchers of 'exceptional talent') *regardless of the contents of their lines of research.*[8]

Early in the planning process, the members of the two committees for the exact sciences and the biological sciences expressed their discontent with CONACYT's view of adopting science policy by means of a planning exercise (Comites 1975). These committees were integrated by Mexico's most prestigious researchers, many of them recipients of the National Prize for the Sciences. Their written commentary can be considered a position paper of the scientific community on the fundamental issues of S&T policy. The main arguments of the committee can be summarized as follows. First, scientific endeavour is described as a creative activity whose main objective is the understanding of nature and whose product is knowledge. This endeavour is capable of having a positive impact on culture and economic development, but these effects are purely accidental. Second, science advances in a linear and accumulative mode, it is also universal in the sense that there is no such thing as a national science. Third, scientific discourse can and must be distinguished from its consequences and from its applications. Thus, science is essentially an academic activity. Technology, on the other hand, is a purely economic application culminating in industrial production. Engineering is the activity bridging these two extremes of the R&D spectrum. Fourth, scientists generate their own replacements in a positive feedback process because they form the new generations of researchers through training and education. Fifth, the only acceptable planning activity of scientific research is 'internal and individual', i.e. every scientist must have the right to decide and select his/her own research priorities and methodologies. As a corollary, the authors of this document concluded that the officials responsible for science policy must 'declare war on one of the most pernicious and dangerous ideas, namely the idea that science can be oriented from the outside'. Science policy has as its main objective, the unrestrictive support, by all means, of scientific endeavour; but it is the scientific community itself that must be responsible for the definition of priorities and modes of research. The allocation of society's resources to specific lines of enquiry must be handed over to scientists. The bottomline argument was that the Mexican state had to increase significantly the financial resources allocated to scientific activities breaking the 1 per cent of GDP barrier. This implied augmenting R&D expenditures by more than 20 per cent per annum (Mexico was allocating 0.5 per cent of GDP to science and technology in 1975). In addition, scientists were to be left alone in their task as sole managers of these resources.

These arguments reveal a naïve perspective of what scientific activity is, of the social responsibility of scientists, and of what science policy is all about. However, the main point here is that the scientific community used this analysis to mark its differences with CONACYT's efforts to define orientations and priorities for R&D. The end result was that the 1976 Indicative Plan for Science and Technology devoted little space to the discussion of basic science, while at the same time explicitly accepted several of the basic guidelines put forth by the Committees on Exact Sciences and on Biological Sciences, particularly in respect to the lack of research priorities in basic research. If the quantitative goals of the S&T Indicative Plan reveal the relative bargaining power of the committees (as well as their power to convince CONACYT of their strategic importance), Table 4.3 shows that basic science does not fare below average. That the plan's quantitative goals for the different sectors are very similar is proof of the incapacity to translate strong priorities in financial terms. For all the pages devoted to analyse the 'style' of scientific and technological dependence, and to the need to change this pervasive pattern, the authors of the indicative plan could not manage to propose significant intersectoral differences in growth rates of R&D expenditures.[9]

The Mexican scientific community is not (and has not been) a monolithic structure. During the debates of the 1960s and 1970s, scientists of recognized prestige openly came out in favour of more state intervention in the allocation of resources and the definition of research priorities. 'Nationalism' was not absent from these positions; in contrast with the views expressed by the basic sciences committees, other researchers (Haro 1967; Lozoya 1973) had already stated that, in general, basic research in Mexico was but a small part of greater research programmes, decided elsewhere (normally in the centres of scientific production disseminated across the developed world). Their conclusion: Mexico had to make a

Table 4.3 Quantitative goals by sector in the 1976 and 1982 indicative plans for science and technology (millions 1975 pesos)

	1976	*1982*	*Annual ROG (%)*
Total	3,107.3	9,278.4	20.0
Basic research	463.0	1,430.9	20.6
Exact and natural sciences	254.8	789.7	20.7
Social sciences	208.2	641.2	20.6
Oriented applied research (OAR)	2,206.2	6,670.8	20.2
Agriculture	621.5	1,688.0	18.1
Fisheries	80.8	284.1	23.3
Manufacturing industries	313.8	1,239.6	25.7
Extractive industries	593.5	1,557.0	17.4
Social welfare	497.2	1,527.4	20.5
Transportation	37.3	186.3	30.7
Other	62.1	188.4	20.3
OAR to national reality	438.1	1,176.7	17.9
Renewable resources	211.3	570.2	17.9
Natural phenomena	62.1	164.7	17.6
Statistics	164.7	441.8	17.8

Note: The resources for R&D in the manufacturing industries were to be channelled by the development bank, Nacional Financiera (NAFIN) through its various programmes. These resources were originated in the federal budget but earmarked for intramural R&D by private firms

Source: Table V, Statistical Appendix 2 (page 363): Plan Nacional Indicativo de Ciencia y Tecnologia. Consejo Nacional de Ciencia y Tecnologia, Mexico, 1976

special effort not only in terms of allocating more financial resources to scientific activity, but also special attention had to be given to the need for redefining its articulation with basic research programmes determined in the scientific centres of the developed world. It is interesting to note that this viewpoint was opposed by both the conservative wing of the scientific community, as well as by members of this community that had been intensely involved in the 1968 movement calling for reforms in the political and bureaucratic apparatus.

The fate of the quantitative goals defined by the Indicative Plan in 1976 was to remain as abstract figures in a piece of paper because the timing of the planning exercise was most unfortunate: by 1976, the presidential race was running its course. The plan would have been adopted by the new administration, had it received the political blessing of the new presidential candidate in the official party. But, again the highly centralized system in Mexico did not lend itself to this kind of relay of plans. Traditionally, plans inherited by a new President from a previous administration are not implemented (in fact, they are substituted as soon as possible). The new President takes power with a new set of plans and government programmes prepared during his campaign; in 1976, as CONACYT was preparing its indicative plan, the presidential campaign of the official party candidate, was quietly preparing a new framework for science and technology policy-making. In 1977, the new President met with the highest scientific hierarchy in Mexico and decided that a new Programme for Science and Technology had to be prepared, formally giving the *coup de grâce* to the indicative plan.

Essentially, the 1978–82 programme describes several dozen specific research projects receiving support from CONACYT (CONACYT 1978). It is a very primitive document, hastily drafted by a few close associates of CONACYT's new Director living under the illusion that Mexico's newly found oil richness would provide the necessary financial stability to increase the amount of resources allocated to scientific research in general. The programme does not discuss quantitative goals. And during those years, CONACYT concentrated on a big scholarship programme designed to form human resources in foreign universities. The programme did not attract any significant opposition from the scientific community for the simple reason that there was nothing in it that was a cause of concern from the point of view of research priorities. The same can be said about the 1984–9 Programme for Science and Technology and its successor for the present administration, the National Programme for Science and Technological Modernization, 1990–4 (CONACYT, 1990) which, as the title suggests, traces a fundamental difference between scientific endeavour and technological activities. It is as if the essence of the 1976 committees of the exact and biological sciences had been adopted, almost literally: science is for the academic sector and technology is for economic activities. Technological development is to be pursued through linkages of university research with private industry,[10] and federally supported R&D to be carried out directly by private firms.

Three main features of science policy during the present administration must be underlined. First, the programme of activities of CONACYT for 1992 explicitly states that it contains no priorities, either in respect to support for scientific activity or for projects in the technological development side of the R&D spectrum (CONACYT 1992). On this crucial point, the Council's position is an extreme and naïve version of the belief in the idea that, in the end, science and technology in general will have a positive 'trickle-down' effect. Second, resources devoted to R&D, both in the realm of basic research and in the area of technological development, have experienced significant growth rates in the past three years. Third, support for research projects is decided only after approval by reviewing committees composed of specialists of the different disciplines. It remains to be seen whether this format is the correct approach to improve the level of technological development. It is also relevant to enquire whether this format is satisfactory for the scientific community, as both the viability and performance of this policy framework crucially depend on this factor.

It would be tempting to conclude that, at last, CONACYT has paid attention to the old requests of the scientific community: increasing resources allocated to R&D without attempting to discriminate between 'high-priority sectors' and non-priority fields would appear to be what the committees dreamt about in 1976. However, there are signs of strained relations between members of the scientific community and CONACYT. To understand this, it may be necessary to observe that between 1981 and 1990 the real salaries of researchers (the majority working in the academic sector) fell by a dramatic 60-70 per cent. In 1984, in order to stem this trend, the federal government set up the System of National Researchers (SNI), which is formally defined by CONACYT as 'a federal programme that supplements the salaries of outstanding researchers'. The researchers must meet the requirements of a tough evaluation committee. Every three or four years (depending on the category) those admitted as 'national researchers' are evaluated and can then either lose or maintain this special fellowship. In addition, every public university has established a system of so-called 'academic stimuli' to supplement a researcher's salary; every six months or every year the recipients must report to internal evaluating committees for renewal. The criteria for renewal in both systems is, essentially, one of productivity: publications (with special emphasis on international refereed journals oriented towards mainstream research programmes), international conferences, seminars, teaching activities, etc. These systems, originally presented as mechanisms designed to reverse the trend of falling salaries, are now being perceived as mechanisms of control that are distorting the way in which research is carried out. Because the salaries of typical academic researchers at the National University today represent not more than 30 per cent of their income (the other 70 per cent is composed of resources from the SNI and the 'stimuli' obtained from the university), it is now said that researchers no longer devote themselves full-time to the university: they are, in effect, employees of the Ministry of Education and of CONACYT (the bodies that administer SNI). From CONACYT's perspective (explicitly outlined by the Council's top officials), researchers should have a *low* base salary, and their income should be supplemented through payoffs to productivity. This, of course, is a powerful irritant for the scientific community. A point of concern for researchers is that these supplementary benefits are not legally considered part and parcel of a salary and they are not part of retirement programmes. Summarizing, although there recently appears to be an increase in federal resources devoted to R&D, a significant part of the increments in R&D expenditures after 1988–9 is explained by the low levels attained after the 1982 debt crisis in Mexico. But most important, the increment is being channelled through the incentives programmes and the SNI. This means that the real salaries of researchers have a long way to go to recover their pre-1982 levels.

Another element having a negative impact on relations with the scientific community is the fact that criteria for academic evaluation and peer reviews have been designed in a centralized manner. In some cases, only international publications are being taken into account as the sole determinant of a researcher's reputation. This discriminates against domestic journals which sometimes struggle on the brink of extinction to obtain good, publishable material. Also, time devoted to students and dissertation guidance is not contemplated as part of performance reports. Finally, researchers have already started to complain that, owing to the competitiveness in science, R&D may be oriented away from creativity and towards goals that may be indifferent to the country's needs.[11]

Although this debate may not lead science policy-making back to the exercise of defining priorities, it may well tend to reduce the bias in favour of big (i.e. international) science. An important indicator here is the series of controversies pitting the National Academy for Scientific Research against CONACYT on issues as diverse as the real salaries of researchers or the sources of R&D spending in Mexico.[12] Mexico's scientific community is starting to react with suspicion and mistrust *vis-à-vis* CONACYT, and increased R&D expenditures are perceived as linked to control and regulation mechanisms. Thus, CONACYT's political role as opposed to its policy functions is seen as dominating the Council's activities. CONACYT may be further than ever before in its history from assuming a leadership role in science policy-making. It is, on the other hand, closer to becoming a body through which the scientific community is subjected to more controls in exchange for additional financial support. The strategic idea that market forces are better at picking out 'winners' has come to the aid of the school of thought that maintains that there is no need to define priorities for R&D activities; but in this process, more pervasive political controls have been established and the scientific community may one day find that things have gone too far.

In the latest attempt to gain more support from the scientific community, a new advisory body for science and technology, directly linked with the executive's office, was created in 1989. Members of this new body, the Consultative Council for the Sciences (CCS), are the winners of the prestigious national science prize (Diario Oficial 1989). Functions of the new organization include channelling all contributions of the scientific community to national development planning; acting as consultant to the executive on scientific matters; and carrying out research on scientific matters. The functions of the CCS are either nonsensical (there is no serious planning exercise) or duplicate CONACYT's powers. Thus, the CCS is nothing but the latest in a long series of attempts of the federal government to appear attentive to the scientific community's advice. In exchange, the CCS helps enrol the political support of part of the community's elite. A lot of time and effort goes into meetings and symposia with the central objective of keeping high-calibre scientists in a position that may help prevent independent criticism.

4 PROTECTIONISM, CONTROLS OF TRANSFER OF TECHNOLOGY AND INDUSTRIAL PROPERTY IN MEXICO: ENTREPRENEURS AND TECHNOCRATS (1945–85). RELATIVE AUTONOMY AND MANIPULATION

Recent literature on industrial policy in Mexico is highly critical of the protectionist framework which has dominated the Mexican economy since the 1940s. A common criticism is that possibilities for deepening the import-substitution strategy became exhausted by the late 1970s and that this was not recognized by the relevant government officials. Thus, Mexico was extremely slow to change gear, to open its economy and move to an export-promotion strategy. This line of argument states that because the size of the domestic market was small, economies of scale were bypassed and, with them, the ability to attain adequate levels of international competitiveness remained beyond reach. In reality, the domestic market was reduced partly owing to grave distortions in the structure of income distribution, patterns of consumption and in regional concentration. Thus, it was not possible to use the domestic market as a solid platform for scale economies and industrial expansion. The potential of the domestic market as a powerful base for industrialization is revealed by research in recent economic history of Mexico: Cárdenas (1987) shows that the base of the domestic market was critical in the very first stages of import substitution (between 1930 and 1939).[13]

Much of today's faith in the allocation powers of the market system is accompanied by a ritual attack on the failed import-substitution strategy of the period 1940–85. The series of criticisms is extended to the use of trade policy instruments, such as tariffs and import permits, as industrial policy instruments. In addition, all the forms of state intervention that coexisted with the protectionist instruments are now condemned as useless, at best, and/or harmful. Today, the extreme view that all forms of state intervention are responsible for the shortcomings of the industrialization process is highly popular with economists.

The dominating objective of political stability, and the incapacity to define long-term objectives for industrial and technological development, as well as the inability to organize different economic forces around these objectives, generated a vacuum where government officials were highly vulnerable to manipulative pressures from private interest groups with a strong inclination for high profitability rates aimed at short-term recovery of investment. The differences in government styles of the different administrations notwithstanding, this fact has been a constant theme in policy implementation since the early days of protectionism until the financial crisis of 1982. An important transformation took place in the 1980s as the economy ceased to be a closed, highly protective environment for domestic firms. This transformation would have been unthinkable a few years before had it not been for the shock waves of the debt crisis. Thus, the financial crisis was simultaneously a crisis of a style of government policy-making that was identified as interventionist.

4.1 Implicit technology policy and the industrial lobby

The political structure that emerged after the first phase of the Mexican revolution (1910–20) proved to be quite a robust and flexible system. There was no single countervailing force capable of checking the power of the national state. But, instead of having a top-heavy scheme for technology policy-making, different interest groups have traditionally played an active role within the state's administrative apparatus and have oriented industrial policy in different (and sometimes contradictory) directions. The best example is provided by the opposing lobbies of manufacturers of final products and intermediate goods in their efforts to manipulate import duties and permits. In this capacity, private interest groups have been able to manipulate different policy instruments without the long-term perspective required by the objectives of science and technology policies.[14] On a large number of important issues, the Mexican experience is very different from the Korean or Taiwanese examples: agriculture has been discriminated against in terms of trade; capital goods industries and high-technology industries were never targeted as strategic industries to be developed; controls on transfer of technology and direct foreign investments, as well as industrial property legislation, are other dimensions where well-marked differences exist. The history of policy-making and of the performance of policy instruments is not a simple one.

Manipulating policies and instruments by interest groups has often coexisted with the relative independence of government officials who have introduced new horizons in policy design and implementation. But, historically, this independent search for alternative initiatives has had to take into account the restrictions imposed by the action of interest groups. This section examines the manipulation of protectionist policy instruments, the introduction of a regulatory system for transfer of technology, and the series of reforms in the industrial property laws of the country. In these three cases, the action of interest groups has coexisted with the deployment of alternative policy initiatives by independent government officials.

Perhaps the best example of policy instrument manipulation by interest groups is provided by the handling of levels of nominal protection through Mexico's tariffs on imports, as well as the inclusion of products in the import permit regime. Tariffs and import permits constituted the key instrument for industrial protection in Mexico between the 1940s and 1985, when Mexico joined the GATT. Until the year 1971, it was thought that Mexico's levels of tariff protection were relatively low and did not exceed 10-15 per cent. This perception was grounded on casual observations of isolated items such as certain non-durable consumer goods and capital goods which were covered by tariffs averaging 8.4 per cent and 10.7 per cent respectively. A study on the structure of nominal effective and implicit effective tariff protection, using nominal levels of protection and the 1960 input–output matrix for the Mexican economy (Bueno 1972) revealed that levels of protection were much higher: the mean nominal tariff protection was 27 per cent. The life-span of these high levels of protection was very long, rendering any argument in terms of 'infant industry' completely irrelevant. In 1985, when Mexico adhered to the General Agreement on Tariffs and Trade (GATT), mean nominal tariff protection was 25 per cent. The high levels of effective protection probably date from as far as back as 1930 for some items, although for other items they had existed since the early 1940s. This means that some industries enjoyed high protection levels for periods exceeding five decades.

Mean nominal effective and mean implicit effective protection were 51 per cent and 46 per cent respectively. This level of protection is clearly underestimated: in addition to tariffs and import permits, imports were charged with an additional 2.5 per cent *ad valorem* earmarked for specific economic objectives (such as export-promotion activities); an additional 3 to 10 per cent established in the Customs Law; and, an additional federal tax of 0.6 per cent. In the case of non-durable consumer goods, nominal effective protection reached 78 per cent, while durable consumer goods showed 67 per cent. The structure of tariff protection revealed a clear bias against protection for the capital goods industries: capital goods had a very low level of tariff protection of only 9 per cent. This structure of protection favoured import substitution of consumer goods and it explicitly demanded disembodied technology (patents and know-how), as well as technology embodied in capital and intermediate goods. It is in this sense that tariffs were one of the most important technology policy instruments in Mexico's recent economic history.

The levels of tariff protection were set up by special committees in which industry representatives met with government officials and examined the cost structure of the industry, as well as the patterns of prices in the international markets. The special committees were co-ordinated by the Ministry of Finance (this was a vestige of the time when these policy instruments were part of fiscal policy). Products subject to import permits were determined by other similar committees co-ordinated by the Ministry of Industry. In both cases, it is possible to say that industry's representatives, particularly in big cities where policy-making is concentrated, were in a good position to increase the number of items covered, as well as the levels of the import duties.

It should be noted that the private sector did not hold a monolithic position *vis-à-vis* protectionism. Normally, producers pressed hard for the highest possible protection for the goods they were producing, and for free importation of the raw materials and intermediate inputs they were using. Producers of radios acted through their lobbies for higher protection in electronic goods and for the lowest possible tariffs on radio parts (the conflict of interest between producers and retail importers was even more pronounced). Already in 1945, during its first national congress, the National Chamber of Manufacturing Industry, CANACINTRA, which represents the interests of small and medium-size industries, condemned in very vigorous terms the position of businessmen importing all sorts of intermediate inputs and consumer goods as enemies of Mexico's industrialization (Izquierdo 1973: 260). Because the small and medium firms affiliated with CANACINTRA did not enjoy easy access to credits and foreign technologies, and they concentrate in manufacturing industries operating with local components, this industrial chamber traditionally acted in favour of protecting the final consumer goods industries. On the other hand, the

Confederation of Industrial Chambers, CONCAMIN, which groups large industry as well as foreign subsidiaries, has traditionally acted in favour of a less protective environment for industry.

Although it may not be possible to establish a quantitative correlation between levels of protection and relative strength of industrialists' lobbies, there are several indicators which help reveal the importance of manipulation. First, many of the benefits originating in the industrial and technology policy instruments were unduly accumulated by single firms. Policy instruments such as the Law for the Promotion of New and Necessary Industries (LINN), and Rule XIV of the General Law of Import, facilitated this high level of redundancy. But it is also plausible that once firms knew how to take advantage of one or two policy instruments, they tended to explore other instruments for new and more important benefits. A study in the mid-1970s (Nadal 1977) found evidence of this redundancy of instruments as firms accumulated advantages from instruments related to tax exemptions and drawbacks, preferential access to credits, duty-free imports of capital goods, etc.[15]

Second, geographical concentration of industrial activity is also important. Between 1930 and 1950, the Federal District saw its participation in the value of manufacturing industry's total production rise from 27 to 38 per cent (Reynolds 1973: 205). Of course, regional location of industry is something that cannot be explained by policy instruments as it is dependent on much more powerful determinants; but in a country where centralism is so intense, proximity to decision-making centres cannot be discarded as another important externality. Firms closer to the ministries where the discretionary power of policy instruments is wielded will have a greater chance of obtaining the additional advantages stemming from tax drawbacks, preferential credits, import licences, higher protection, etc.

Third, most of the policy instruments with benefits accruing to firms were managed by small units in the relevant ministries where officials had a high degree of discretionary power in everyday decision-making. There was little or no co-ordination between these different units.[16] Discretionary power was important in so far as it covered decisions related to eligibility of companies for many of these policies' advantages, as well as the levels of benefits accruing to the firms (percentages of tax redeemed through credits and drawbacks, amounts of credits in preferential loans, interest on loans and payback periods, etc.).[17]

The power of the protectionist lobby was such that even after the presidential decision of 1977–8 to join the GATT, and after negotiations had taken place and a Protocol had been agreed between Mexico and the GATT's contracting parties in 1979, the final step could not be carried out.

Repudiation of the protectionist strategy and change to an open-economy approach was only possible after the major crisis of 1982 shook the Mexican system. In August of that year, the Mexican government declared a unilateral moratorium on its service of the external debt. This event ignited a major political restructuring; the political forces within government that were committed to opening the economy gained the upper hand, as the international financial community and the obligations acquired *vis-à-vis* the International Monetary Fund forced the Mexican government to enter into a macroeconomic adjustment process, to deregulate industry and services, and to shift to a more open economy. This restructuring allowed the Government to start renegotiating adherence to the organization with GATT's Secretariat in order to implement the 1979 Protocol. The adherence process culminated in 1985.

The pressures of international agents have also played a role, but the lesson of this analysis is that they are important only when the internal conditions are ready to accept them as allies. Two examples, one where foreign pressures were resisted, and another where the government succumbed to foreign pressure, are relevant here. The first refers to the case of the Registry of Transfer of Technology (ROTT) established in 1972. UNCTAD gave full support; but Council of the Americas and other US groups were totally opposed. In spite of these pressures, why was the ROTT implemented? The other example is the recently enacted Law for the Promotion of Intellectual Property, rushed through an ignorant Congress in 1991 as a gesture to the American government in order to obtain the fast-track authorization for the negotiation of NAFTA. Macroeconomic interests prevailed over sectoral interests (and it was impossible for the Chamber of the Pharmaceutical Industry to change this, regardless of its relative importance in the context of transfer of alternative process technologies). In this last instance, international agents were totally helpless against the interventionist policies of this same ministry a decade and a half before. Why were they successful in their attempt to radically transform the intellectual property law? The next two sections will try to throw some light on these questions.

4.2 Explicit technology policy and the relative autonomy of government officials

In 1972 the Mexican government enacted a law requiring all contracts on transfer of technology to be registered in a special office in the Ministry of Industry. Noncompliance with this requirement was accompanied by nullification of all legal effects of the corresponding agreements. The new law covered agreements on patents, trademarks, unpatented know-how, technical assistance, engineering (basic and detailed) services, training of technical personnel and management services. The parties to these agreements had 60 days after signing to register their contracts in the Registry of Transfer of Technology (ROTT).

In essence, the purpose of this requirement was to open licensing agreements to scrutiny by government experts in order to verify that restrictive clauses harmful for the economy would not be included as part of these agreements. The forbidden restrictive clauses can be grouped in five different categories:

1 excessive price;
2 excessive duration of agreements, in no case can duration exceed ten years;
3 transfer of freely available (identical) technology in Mexico;
4 restrictions or limitations on exports, use of supplementary technologies and purchases of equipment, management, R&D, use of certain inputs, cession of patents and trademarks or unpatented know-how;
5 submission to foreign courts in cases of conflicts of interpretation and/or fulfilment of the agreements' terms.

As an important concession to the Registry's opposition, exceptions were allowed (granted through the discretionary administration of the Registry's law and regulations), but not in the case of above categories 2, 3, 5 and several items of category 4.

When the ROTT was established in 1972 no information was available on the number of technology agreements existing in Mexico. Already existing contracts were given a two-year period to conform to the new law, but they were to be registered for information purposes. During the first three months after the law was enacted 5,625 contracts were presented to the ROTT: 4,112 were to be renegotiated during the two-year grace period, and 1,513 were submitted for normal registration. By mid-1975 the total number of contracts rose to 6,497, making the Registry by far the most important technology policy instrument in Mexico. Over the next years, the ROTT was responsible for improving the bargaining position of recipient firms, as well as for significantly reducing royalty payments to foreign licensors.

Opposition to the Registry was spearheaded by Mexican firms which perceived this instrument as an obstacle to their access to foreign sources of technology. Although a significant effort was carried out (essentially through CONCAMIN, and the Confederation of Entrepreneurs, COPARMEX) to convince the representatives of the private sector, nevertheless domestic firms were, in general, more inclined to be against the

ROTT. The National Chamber of Manufacturing Industry, CANACINTRA, which groups medium and smaller-sized firms in the manufacturing sector, and which has traditionally supported nationalistic policy measures such as the nationalization of the electrical industry, was also an important agent in this process. Although CANACINTRA was less vocal in its opposition to the ROTT, it nevertheless was not wholeheartedly in favour of this new instrument. This was perhaps the result of internal divisions within this organism.

The negative reaction of the Mexican private sector was probably due to several factors. Any initiative to increase the state's regulatory role would have provoked some degree of opposition. But the very intrusive powers of the ROTT over what was traditionally (and still is) considered a very intimate and essential part of a firm's operations was a formidable irritant. It is also possible that the establishment of the ROTT was perceived as a first step in the direction of regulating unfair business practices and eliminating the negative effects of monopolistic industrial structures. The intrusive nature of the ROTT was sometimes defended by the Ministry of Industry by comparing it to policies already existing in highly industrialized capitalist economies. The Sherman Anti-Trust Act in the United States was often mentioned in this context, as an example of a policy that sought to eliminate restrictive business practices emanating from cross-licensing agreements designed to carve markets in a monopolistic framework, or certain restrictive clauses in patent and trademarks licensing agreements. However, this line of defence backfired because it did not assuage the anxieties of the private sector, but rather intensified their worst fears about the interventionist and regulatory tendency of the administration. In this context, the ROTT was seen only as the first step of a sequence of highly regulatory policies.

In the international scene it is interesting to observe that the new law detonated a vicious offensive by such organizations as the Council of the Americas. But, in reality, the multinational corporations (MNCs) already operating in Mexico did not have a strong negative reaction. Of course, they would have preferred to continue operations in a less regulatory environment. On the other hand, they were not particularly affected by the restrictions imposed by the ROTT. This is due to the fact that the head firm of an MNC does not require the inclusion of restrictive clauses in a technology agreement. It can impose or dictate restrictive behaviour by the sole reason that the wholly owned subsidiary does not have alternative options. Besides, one of the main advantages that MNCs had during the 1970s was not affected by the ROTT: fiscal treatment of royalty payments (associated to intrafirm transfer of technology) was more favourable than fiscal treatment of profit remittances to the head firm. This was unchanged by the Registry, although (as in the case of domestic firms) MNCs were extremely suspicious about the intrusiveness of this policy in this most reserved domain of a firm's operations.[18]

In addition, the ROTT was seen as an obstacle to obtaining technology in an already difficult market. Efforts to sell the idea of the ROTT as an instrument that would enhance the negotiating power of entrepreneurs in their acquisition of technology were not very successful. It is interesting to note that in this context, entrepreneurs reacted almost parallel to the scientific community's views on research priorities and additional R&D resources: entrepreneurs thought of themselves as the ideal agents for the choice of core and periphery technologies, as well as the best-placed agents to understand all issues related to direct (royalties, lump-sum payments) and indirect (arising from restrictive clauses) costs. From this perspective, entrepreneurs concluded that any form of state intervention would be detrimental to transfer of technology operations.

As with other S&T policy instruments, the ROTT was subjected to the pressures of existing administrative structures and political arrangements.[19] The fact that its administration was endowed with discretionary powers only exacerbated the intensity with which these pressures were exercised. The ROTT was particularly vulnerable to pressures from the giant state-owned corporations such as PEMEX and the Federal Electricity Commission. These entities probably showed more contempt for the ROTT's functions than did most private firms. The technology agreements of PEMEX and CFE included many of the restrictive clauses forbidden by the ROTT. In addition, many agreements did not even contain an explicit determination of the amount of royalty to be charged; thus, the golden rule of the Registry, which was to have all agreements clearly spell out the direct costs of the operation, was broken by these agreements. Nevertheless, the ROTT succumbed to the pressure and had to grant its approval on agreements entered by these giant public corporations even though they violated the law where no exceptions were to be accorded.

The ROTT was created in almost complete isolation of the private sector's pressures and of the institutional constraints of the public sector. When the financial crisis of 1982 struck, political cost was also laid on the regulatory side of the state's actions; together with the Foreign Investment Commission, the ROTT was virtually condemned to a slow but effective death.[20] Summarizing, the private sector had to endure (between 1973–84) what it considered an invasion of its prerogatives; it could not muster enough leverage to change this law until 1983–4, when a new administration started implementing in an extremely 'flexible' way the main disposition of the ROTT's law. For all practical purposes, the ROTT became ineffective during the mid-1980s. The public sector agencies, on the other hand, were luckier in their attempts to escape the Registry's regulatory functions.

4.3 Explicit technology policy, relative autonomy of technocrats and the imperatives of the external sector: the case of industrial property

On the promotional side of technology policies, the industrial property system is one of the most important elements. Mexico has been a member of the World Intellectual Property Organization (WIPO) since its origins, having ratified the relevant international conventions on patents, trademarks and copyrights. The law suffered two drastically opposing reforms in 1976 and 1991. The analysis of how these reforms came about reveals once again that government officials operate with a considerable degree of autonomy *vis-à-vis* domestic interest groups. But they exhibit considerable less freedom of action when confronted with powerful international pressure and when domestic macroeconomic priorities dominate long-term technology policies.

In 1976 a new law for patents and trademarks was enacted by the federal Congress. Among the changes with respect to the old legislation (dating from 1942) were the following:

1 Chemical and pharmaceutical products, antipollution devices and all innovations related to nuclear energy were excluded from the patent system, but they could be protected by inventor's certificates; it must be added that pharmaceutical products would be subject to patents starting in 1997.
2 Duration of patents was reduced from 15 to 10 years.
3 Compulsory licences could be demanded when the patent remained unexploited during a period of three years after the patent was granted (compulsory licences could be demanded if economic exploitation was interrupted for more than six months or if production was not enough to cover the domestic market).
4 All trademarks had to be exploited in order to remain under the protection of the law.
5 All foreign trademarks covering products manufactured in Mexico were to be used linked to trademarks originally registered in Mexico (the Mexican trademark was to be used in equally ostensible patterns and it could not consist of foreign words).[21]

This law was drafted by a small group of officials in the Ministry of Industry. Consultations with the private sector were kept to a minimum. It was well known that the local Chamber of the Pharmaceutical Industry, CANIFARMA, was split down the middle over the critical issue of patentability of pharmaceutical products, with the international component of the chamber opposing the new law on this issue and requesting by all possible means the extension of patent coverage to pharmaceutical products. The small- and medium-firm component of the National Association of the Chemical Industries, ANIQ, was in favour of the law, forming a unified block together with half of the firms affiliated with CANIFARMA. The prevention of patents on chemical and pharmaceutical products was designed to enhance independent development of process technologies, and this was an interesting measure for domestic firms envisaging joint ventures (particularly with Italian firms, since Italy did not recognize patents on pharmaceutical products). But subsidiaries of MNCs remained adamant in their opposition to this part of the law.

It was the section on linkages of trademarks that brought about the fiercest international reaction. In a memorandum from the United States Trademark Association addressed to the Secretary of State, Henry Kissinger, the terms of this opposition were clearly spelt out: '[the articles on linkages of Mexican trademarks with foreign trademarks] are clearly in violation of the Paris Convention which established that foreigners will have the same rights as nationals' (USTA 1976). As a result, the swift and vigorous intervention of the then US Secretary of State was requested to modify this law in order to prevent this violation of 'the most elementary principles of honest competition'.

In 1977 the new administration of José López Portillo was sworn to power, ushered under the extremely optimistic auspices of what later became known as the oil boom. For the next few years economic policy-making was dominated by a mixture of nationalistic pride and careless mismanagement of the country's natural and financial resources. These years were not favourable to attempts to change the terms of the laws on transfer of technology or on patents and trademarks. Opposition from the traditional international and domestic quarters had to wait for better times. These came along unexpectedly after the 1982 crisis, which was accompanied by the idea that the regulatory measures of the previous two administrations were the real culprits of bad economic periods.

By 1983–4, under the administration of Miguel de la Madrid, the private sector, both domestic and international, was pressing hard for economic reform in Mexico. Between 1983–8, the regulatory environment started to change gradually, with perhaps the biggest single event on this front being Mexico's adherence to GATT in 1985. The main political cost attributed to the regulatory measures of the 1970s would later be exacted on several fronts.

The Salinas administration, tainted by the worst scandal of fraudulent elections, took power in late 1988.[22] In 1991, Mexicans learned via the international press that their government was starting negotiations with the United States in order to reach a free-trade agreement. The US executive branch announced it would request a so-called fast-track authorization from Congress to negotiate the free-trade agreement with Mexico. But in itself, this was presented as a difficult negotiation between the two branches of government: opposition was strong in the US Congress dominated by the protectionist democrats, and even before formal negotiations were engaged for the free-trade agreement, Mexico (the demander of the agreement) had to demonstrate it was negotiating in good faith.

By definition, no sectoral consideration can outweigh priorities defined in terms of macroeconomic variables. No coalition of private interest groups, however powerful it may be, will be able to launch successful policy initiatives as long as the basic tenet of faith in the allocation powers of the market is held, since a corollary of this tenet is that the market cannot operate efficiently without stability of macroeconomic variables. So, sectoral priorities must yield to macroeconomic objectives. NAFTA became in the present administration one of the key instruments of economic policy, both because it is an instrument to attract foreign investment in significant amounts, enabling the economy adequately to finance a growing current deficit (itself the consequence of the abrupt opening of the economy), and because it was assumed an international treaty would 'lock in' the economic reforms of the administration. The locking in of the administration's sweeping reforms was deemed a must as it would allow foreign investors to see Mexico as providing a stable environment in the long term, meeting the most stringent expectations of foreign investors. Thus, Mexico would become a competitive environment in the international struggle for capital resources.[23] Government officials considered the attainment of the fast-track authorization from the US Congress an utmost priority and no amount of sectoral lobbying would change that. A new patent law was not perceived as a sacrifice.

Consequently, the patent and trademark legislation was radically transformed in June 1991. The main components in the new law were the following:

1 Chemical and pharmaceutical products and processes, antipollution devices and all innovations related to biotechnologies were covered by patent protection.
2 Duration of patents was increased to 20 years.
3 Compulsory licenses almost completely disappeared.

4 Requirements on exploitation of trademarks were relaxed.
5 Linkages of Mexican and international trademarks disappeared.

It is ironic that the same administration that embraced the dogma of free-market economics and deregulation was to introduce the longest lifetime for monopolies (i.e. patents) that Mexico has ever seen.[24] The new orientation was so markedly in favour of protecting patents that, in some cases, the law contradicted even the most vital elements of patent law. For example, in the case of pharmaceutical products and processes, duration of patents may be extended. And, more surprising is the fact that for already exploited pharmaceutical products, patent applications would be accepted. This totally contradicts the standard industrial property requirement of novelty for patent applications. Thus, even the foremost requirement of any patent legislation, and that which defines its rationale, novelty of the protected invention, was bypassed by the new law in the case of pharmaceutical products.

The new law was approved by Congress without debate. Opposition parties, both to the left and right of the official party, did not even attempt a cursory examination of the new law. The Ministry of Industry and Trade was *de facto* sole legislator in this process. But part of the domestic private sector in the pharmaceutical industry protested through the normal institutional channels (i.e. the CANIFARMA acted as the main lobbyist with the Ministry of Industry and Trade). The main issue of the dispute was patentability of pharmaceutical products. Several national firms had entered into agreements with Italian companies and succeeded in developing alternative processes to highly profitable products initially produced by the powerful Swiss pharmaceutical industry. If pharmaceutical products were left without patent protection, any firm could enter into an agreement for alternative processes leading to unprotected (and profitable) products. The powerful international lobby of Swiss, American, French and German pharmaceutical multinationals were not ready to leave this door unclosed.

Various official statements emphasized the need to harmonize Mexico's industrial property system with the requirements of the international patent system and, in particular, with the Paris Convention. It was stated that as part of becoming an open economy, Mexico had to recognize the international imperatives of the patent system and act accordingly. No mention was made of the fact that the lifetime of patents and the scope of patentable inventions is not covered by the Paris Convention. This gross misrepresentation of what it means to be a party to the Paris Convention was maliciously used to justify the need for a new law with long patent lifetimes and expanded patentability.

This analysis may explain why, now that this strong patent system has been set up, there is no real interest in realizing all the potential this instrument offers for a sound technology policy (both as a promotional tool for technological development, as patents are in theory considered to become, and as a source of technical information aiding the diffusion of innovations in the economy through a well-organized market of innovations). The implementing regulations of the new law have not been prepared and there is a backlog of more than 21,000 pending patent applications. The new law includes the creation of the Mexican Industrial Property Institute (MIPI) responsible for the management of the entire system; this entity could act as a powerful disseminator of technical information. Once again the predominance of macroeconomic objectives is clear: it appears that the administration does not favour the implementation of this aspect of the law as it would require important budgetary allocations and thus go against the objectives of reducing public expenditures. In fact, the institute could be a self-financing entity with resources coming from fees charged to applicants, patent holders and trademark owners.

The Mexican patent office still has a very small number of patent examiners charged with the inspection of patent applications: there are 12 examiners in the electromechanical department, and 13 examiners in the chemical department. In comparison, Brazil has approximately 80 examiners and Korea has more than 400 (and a plan to reach the figure of 500 examiners by 1995); the US Patent Office has more than 2,000 examiners. The resources allocated to the management of the Patent Office are clearly inadequate. However, the Mexican Patent Office has already entered into several collaboration agreements with the US Patent Office. Applications are partially examined by the US Patent Office; because the novelty examination is crucial in the decision-making process, this practice could have far-reaching implications. Patent granting decisions taken by the US Patent Office on behalf of the Mexican Patent Office may not be impartial. The US already owns the vast majority of patents in Mexico, as shown in Table 4.4. This is particularly important in the context of NAFTA.

5 FINAL COMMENTS: THE NEW S&T POLICY IN MEXICO

Mexico's industrial economic growth between 1940 and 1980 emphasized an inward-looking style of industrialization. Relations between producers of final goods and their clients, as well as their suppliers, were not good. Suppliers were not well treated by end producers, and in times of economic stress they were abandoned. The costs of re-establishing a network of suppliers were not considered important. Mexican producers (as well as

Table 4.4 Patents granted by nationality, Mexican Patent Office, 1980–91

Year	*Mexico*	*Germany*	*Japan*	*USA*	*Other*	*Total*
1980	165	176	55	1,140	460	1,996
1981	188	168	59	1,225	570	2,210
1982	197	170	88	1,524	1,174	2,583
1983	162	175	101	1,222	587	2,247
1984	138	109	88	981	421	1,737
1985	100	85	52	646	289	1,172
1986	41	73	43	605	225	987
1987	67	78	69	625	317	1,156
1988	256	229	186	1,697	790	3,158
1989	194	156	84	1,237	470	2,141
1990	132	111	72	957	348	1,620
1991	129	95	67	801	268	1,360
TOTAL	1,769	1,625	964	12,660	5,919	22,367

Source: Science and Technology Council, Indicators, 1992

foreign investors operating in Mexico) operated with the sense that whatever reached the shelves of retail stores would be sold, irrespective of quality considerations. They produced for an urban market in an economy with a highly skewed income distribution structure. In addition, the pattern of labour relations was also extremely archaic: when demand went down, workers were fired, and when times of bonanza came, they were hired. Freedom of association and the right to strike were consecrated in the relevant legislation, but were not respected in practice. Training and all forms of investment in human resources were seen as a necessary evil to be minimized and, if possible, dispensed with altogether. This reinforced the sense that a new pattern of labour relations was not needed. Finally, firms operated with a very rigid internal structure in which design departments functioned in isolation from production, finance and marketing departments. The Taylorist paradigm dominated the workshop, and to a large extent, continues to do so today.

Advocates of opening the economy, and of NAFTA, have repeatedly stated that producing for the international market will provide access to scale economies which were not available due to the closed nature of the previous growth model.[25] This crass view ignores the most basic facts of how firms and whole industries compete, survive or die in the contemporary arena of international trade. Economies of scale alone are not the main source of international competitiveness in a world where markets have become increasingly differentiated and where catering to specific niches with highly differentiated products has become the key to profitable positioning of industries. If the Mexican industrial establishment is to compete and survive in this context, which emerged during the 1970s as a matter of a number of factors, a whole new system of production will have to be adopted. This new system will require a more attentive attitude towards the evolving needs of clients and final markets. It must consider the requirements of suppliers. In particular, producers will have to integrate the launching of new products with future investment plans of suppliers; this requires not only advance planning, but simultaneous engineering where the products involved are sufficiently complex. This implies radically new technological relations between firms situated within the fabric of vertically integrated sectors.

The new industrial system will also necessitate the introduction of new labour relations. Because trade unions were organized around the corporatist structure of the political apparatus, they have not been occupied with the objectives of training the workforce and establishing a new set of relations between workers and firms. This must change radically in the short term if greater international competitiveness is to be attained. On the other hand, it is not sufficient to link productivity with the evolution of wages (as has been recently hailed as great success from NAFTA's parallel agreement on labour) unless labour relations undergo a major overhaul.[26] Competitiveness cannot be grounded on low wages alone. The evolution of productivity is not only determined by the intensity of a worker's activities; it is also critically determined by a host of other factors, including the product's design, shopfloor layout and, above all other considerations, training and education of the workforce. As long as this is not understood, productivity, and with it, competitiveness, will not be afforded adequate attention.

The sources of competitiveness are multivariate and require an adequate policy framework. They integrate a set of new requirements of S&T policy. So, the question to weigh when looking at S&T policy in an open economy is not whether there is room for it, but rather what are the requirements that it must satisfy. Given the role of technology in international trade and as base for internal capital accumulation, science and technology policies are an important and even a necessary asset for an open economy. There are two emerging questions in this context. First, who are the relevant actors that can launch new and intelligent science policy initiatives? Second, are S&T policy requirements in today's economy different or the same as the requirements of the Mexican economy in previous decades?

The current administration has emphasized time and again that government officials are bad players in the game of choosing economic winners and that this task should be the sole responsibility of the market mechanism. But, it still remains to be seen if the same interest groups that manipulated policy instruments in the past will remain quietly on the sidelines as the market mechanism performs its allocation chores. If the recently concluded NAFTA negotiations are an indicator of which attitude will prevail, it would appear that powerful interest groups will not adopt a passive attitude. But the stakes have changed and the name of the game may be different: new links with suppliers and clients, as well as with competitors; new relations with workers and middle level technicians; and new structures inside firms so that design capabilities are fully integrated with production activities, quality control and marketing, to provide the frame of reference for new technology policy requirements. It would be dangerous to ignore the vital importance of these requirements. If they do not become a feature of the industrial apparatus, the full potential of an export-based strategy for growth may not be realized.

As in the past, the federal government may not be willing (or capable) of imposing well-defined sectoral priorities in order to co-ordinate financial, technological and marketing operations of large undertakings with the potential to generate stable, competitive advantages for a new growth pattern. Instead, the government is betting on how private firms will reorganize themselves, either through strategic alliances and market restructuring, or through the consequences of more painful and costly predator behaviour – foreclosures and bankruptcies. Any other form of government intervention is considered to be protectionist in nature and distortive of relative price structures.

Thus, government officials have so far closed the door of access to government policies that could help in the transition to a new industrial development paradigm. This myopic vision neglects how markets and policies interact, especially in the context of highly open and deregulated economies. Recent work on industrial and technology policies in the United States (MIT 1989; Tyson 1992) clearly reveals not only that much room for manoeuvre is available, but that when open economies do not take advantage of this, serious losses will be incurred in terms of international competitiveness.

The federal government is relying on maintaining macroeconomic stability, particularly in relation to public spending, the rate of exchange and monetary variables. Aside from this, sectoral policies and S&T policy will continue to occupy a secondary position. However, competitiveness in the international environment may require more than just providing for this macroeconomic stability. But whatever the technology and industrial policy to emerge, a fundamental condition is that it must be periodically revised. Domestic industries must be subjected to the test of periodic revisions in order to demonstrate that the results of new forms of government intervention are being well directed and are not becoming a mechanism for the artificial protection of quasi-rents.

The recent evolution of industrial policy-making in Mexico shows that the big industrial-financial groups in Mexico have been well positioned to defend and promote their interests. This is clear from the experience of the NAFTA negotiations, when big industry was able to follow and monitor the process closely (and even take a leading role in key decisions regarding the calendar and pace of tariff reductions). These big industrial groups will be in a position to enter strategic alliances, or joint ventures for risk-sharing and market penetration. But what about the rest of the industrial fabric in Mexico? The NAFTA negotiations point out that big industry's interests are well taken care of, but treatment of small and medium industries does not receive the same level of attention. In the future, these firms must be focused on if adequate manufacturing networks are to be established. It is important that these firms rise to the occasion by showing in more articulate ways how their own *specific* interests coincide with the *general* interests of international competitiveness.

A final word on the role of international factors is called for. Invoking international factors as constraints and advantages is typical of government officials these days. For example, just as in the case of the industrial property system, where the new law was justified through arguments related to the nature of the commitments of the parties to the Paris Convention, CONACYT is justifying its 'no priorities' dogma in science policy as consistent with GATT's rules on subsidies. This preposterous claim is not valid because the GATT's Code on Subsidies does not rule out support of upstream R&D activities (i.e. subsidies for upstream R&D do not serve to justify detonating countervailing measures under accepted trade remedy legislation). And it is very difficult and time-consuming to prove that technology development is so close to production that it constitutes the basis for unfair competition. Finally, GATT's rules on subsidies do not impede subsidies for activities unrelated to international trade. Erroneously or mischievously assuming that this is so, will close important strategic options for policy-makers in Mexico.

The recent controversy between the scientific community and CONACYT regarding the share of the private sector in total R&D expenditure may indicate a fracture between these two opposing forces. However, we must not expect that the scientific community will be an important actor in demanding that science policy formulation be carried out in terms other than through the allocation powers of the market mechanism. The idea that market forces should take care of allocation patterns may cause some discomfort within the scientific community because basic research will tend to suffer from this. However, for obvious reasons, the non-interventionist side of the picture will remain quite appealing to the scientific community working in the academic sector.

Summarizing, first, there is a need for science and technology policy. Competitive advantages are neither inherited nor even a function of factor endowments; they are generated and moulded by S&T policy. Second, becoming an open economy does not entail renouncing S&T policy. On the contrary, the case can be made that open economies require a well-defined S&T policy. Third, rule of law and accountability are important to help S&T policy maintain its objectives. These two factors may make S&T policy less vulnerable to manipulation by interest groups. Another feature of governance that may prevent S&T policy (and government policies in general) from becoming distorted and simply benefit private interest groups, is the existence of a professional legislative branch that exercises its prerogatives in a truly independent manner. This last point brings us to the issue of the relationship between forms of government and S&T policy. There does not appear to be a direct relation between democracy and a successful science and technology policy. However, some elements of democratic forms of political life, such as rule of law and accountability (checks and balances) are important to keep sectoral policies in line with their original objectives. In addition, it may be appropriate to conclude this chapter by stating that the question of democracy is far too important to be justified by the needs of a successful S&T policy. Although deregulation of the economy has proceeded at an accelerated pace, deregulating the market for political freedom in Mexico has a long way to go. Regardless of its implications for the design and implementation of sectoral policies, a true democratic form of government remains the most important objective for Mexico.

NOTES

1 Science and Technology Programme, Centre for Economic Studies, El Colegio de Mexico.

2 These issues have attracted the attention of researchers in the past, but they deserve more attention. An early analysis of how institutional structures condition the effectiveness of science and technology policy initiatives is found in Turckan (1971). This study of TUBITAK, the Science and Technology Council in Turkey, concluded that because of the rigidity of the administrative structure, TUBITAK could not carry out its co-ordination tasks and ended up reinforcing the state of things it was originally designed to change.

3 According to data released by UNESCO, the private sector's share of R&D in other countries is as follows: Brazil 19.8 per cent; Chile 18.2 per cent; Greece 19.2 per cent, Portugal 27.4 per cent.

4 One of the most important examples of how international actors may influence entire research trajectories is provided by the role of the Rockefeller Foundation setting up, in the 1940s, the National Institute for Agricultural Research (INIA). The INIA was the result of the merger of the Office of Special Studies (created as a result of the agreement between the Rockefeller Foundation and the Mexican government and the Agricultural Research Institute. The focus of INIA was to attack the problems of low productivity in agriculture through high-yield varieties and input intensive technological systems. Thus, attention to the local cultivars was relegated to a secondary task. The development of high-yield varieties (HYVs) did increase productivity, but it also increased inequalities, environmental problems and resulted in increased genetic erosion, a major problem for the future of food production. The end result was a reorientation of agricultural R&D to commercial and input intensive agriculture.

5 The *habeas corpus* legal recourse is not sufficient because it simply prevents the execution of an administrative act or of a judicial decision if there is enough evidence that it violates the due process of law or one of the individual fundamental human rights. This recourse cannot be used in cases where an individual or group of individuals considers a government department has violated the law.

6 The first organism responsible for science and technology policy-making in Mexico was the National Council for Higher Education and Scientific Research (CONESIC) which was founded in 1935. This body was active between 1935 and 1938, trying to promote scientific research in various ways, but its mandate was not clear and a serious dispersal of efforts ensued. Its main contribution was the advisory role it played in the foundation of the National Polytechnic Institute. In 1942 the Commission for the Promotion and Co-ordination of Scientific Research (CICIC) was established and charged with promoting research in the fields of mathematics, physics, chemistry and biology. It had a very limited budget and its impact was negligible. In 1950 this body was substituted by the National Institute for Scientific Research (INIC) with an almost identical mandate, but greater institutional stability. In 1961 INIC suffered several important reforms, effectively inaugurating the presence of formal science and technology policy-making in Mexico. With these reforms, INIC became involved in activities such as the promotion of linkages between research centres and laboratories with private and public enterprises. However, its main function was still centred in training human resources and managing a fellowship programme.

7 The autonomy of the universities is guaranteed by federal law and it is considered a fundamental feature of the higher education system. Autonomy is considered to be the guarantee for freedom of speech inside the universities. Autonomy for the National University was obtained through a political movement in 1929 which mobilized thousands of teachers and students. In the years that followed, all the public universities in Mexico were accorded this same status. The granting of this peculiar status by the federal government coincided with the reorganization of three previously independent research centres as part

of the National University: the National Astronomical Observatory, the National Geological Institute and the Direction of Biological Research.

8 Nothing of this sort appears in the chapter on basic research in the social sciences, where an explicit research agenda is presented as the Council's list of priorities (ibid.: 129–30).

9 The intrasectoral differences can be explained in terms of the very low starting points (i.e. transportation) or the size of the subsector (manufacturing industries).

10 Foremost in the list of mechanisms linking academic research with firms in the public and private sectors are university offices acting as intermediaries between laboratories and firms. The two largest universities in Mexico, UNAM and UAM, established these services in the mid-1980s with mixed results. It is quite possible that these linkages will become more frequent in the near future. It is still too early to assess the performance of these mechanisms; together with some success stories, many failures have been reported. In addition, units charged with enhancing linkages with private firms have also been set up in some government laboratories.

11 The weariness of the international scientific community with the competitive character of science today (and with the system of peer review) has already made its appearance in the editorial pages of prestigious scientific journals. See, for example, Maddox (1993).

12 Mexico has requested formal entry to the Organization for Economic Co-operation and Development (OECD). CONACYT's release of information regarding R&D spending patterns by the private sector, according to which the private sector accounts for up to 22 per cent of total R&D spending in Mexico, has been met with scepticism, particularly from researchers (and the National Academy). In fact, it has been seen as part of a public relations campaign orchestrated by CONACYT to present an image similar to some OECD countries in so far as R&D is concerned. The important point here is that, whether this figure is accurate or not, the debate has sometimes taken a bitter tone, injecting distrust into the relations between the scientific community and CONACYT.

13 The early import-substitution process was also based on the evolution of terms of trade and relative prices of imported versus domestic goods. However, the dynamics of domestic demand was the main factor behind the import-substitution process of that decade.

14 In Korea, interest groups have also benefited from the orientation of industrial and technology policy. But their short-term interests were always subordinated to long-term objectives imposed by a strong administration.

15 In some cases, subsidies in terms of lower energy prices, water, petrochemical inputs, preferential transport tariffs through state-run railway system, soft loans for purchases of land, etc., must be added to the list of advantages accumulated by individual firms. This is particularly true in the case of the larger and better organized firms.

16 The existence of several 'co-ordinating commissions' did not change this situation. For example, the Co-ordinating Commission for Industrial Policy in the Public Sector (set up in the mid-1970s) was born helpless. The different ministries, state-controlled firms and other state entities (PEMEX, Federal Electricity Commission, etc.) were powerful agents which protected with zeal

their areas of influence. Instead of risking detonating costly turf battles, non-aggression pacts were implicitly agreed upon, and the co-ordinating functions of these commissions were never carried out.

17 Discretionary instruments were the policy response to the traditional dilemma of using self-enforcing policy instruments which equally affect all economic agents, or more discrete instruments which discriminate on the basis of criteria such as nationality of stockholders, contribution to employment and regional development, and, of course, level of integration with local industry.

18 Royalty payments continue to weigh heavily in total remittances to sources of direct foreign investment. According to official data, royalty payments represent between 25 and 33 per cent of total remittances to sources of direct foreign investment. To this date, the fiscal treatment of both royalties and profit remittances is the same, so the incentive that existed before has disappeared.

19 The ROTT was the instrument of a non-discriminatory policy to increase the bargaining power of recipient firms in technology agreements. This was aimed at reducing the direct and indirect (i.e. through restrictive clauses) costs of licensing agreements. In this sense, the ROTT was a policy instrument affecting the formal conditions of a contractual relation. The ROTT did not affect or try to influence the substantive decisions related to sectors or types of technology.

20 The Foreign Investment Commission (FIC) was established in 1973. This body was responsible for implementing the law on foreign investment which, among other things, limited the participation of foreign investors to 49 per cent of equity of any given firm. In several sectors foreign investment was admitted up to 40 per cent (autoparts, secondary petrochemicals) and 34 per cent (mining). Foreign investment was not permitted in oil, transportation, radio and television networks, forest products, etc. The FIC was the body with important discretionary powers in charge of negotiating and authorizing the different levels of participation of foreign capital. In this task it used a set of heterogeneous criteria relative to regional development, employment, exports, technological development, etc. During the 1970s, the FIC was the companion instrument of the ROTT.

21 In practical terms, this means that the law created the legal entity of a 'Mexican trademark'. The rationale for this was as follows: in licensing agreements where trademarks are involved, even when patented and unpatented technology can be assimilated by the licensee and thus eliminate the need for a renewal of the original licensing agreement, the ownership of the trademark and brandnames by the licensor has often been the tool through which a market is controlled. Thus, a new licensing agreement is required, even though the licensee has become self-reliant from the point of view of the technology. As trademarks have longer lifespans than patents, this control over trademarks becomes of critical importance. To help Mexican firms break this bondage, the law introduced the linkage of trademarks so that, when licensing agreements expired, consumers could still identify the trademark or brand of the national firm as having been associated with the foreign trademark and brandname. A more competitive firm would have been created, not only from the point of view of the process and product technology, but also from the point of view of marketing.

22 In August 1988 new presidential elections took place. After decades of corruption and economic disaster, the governing party was punished at the polls by a coalition of disenchanted peasants, workers and middle-class urban groups who

had seen their incomes drastically eroded. For the first time in decades, the electoral process revealed very strong trends favouring the opposition parties and the official machinery controlling the balloting resorted to fraud. The precise data of the balloting may never be known (as the ballots themselves were incinerated in 1991 by order of government officials), but the Salinas administration will never be able to erase the impression of illegitimacy and illegality surrounding its coming into power.

23 The shift of eastern European countries away from the Soviet sphere of influence was perceived in Latin America, Africa and Asia as the creation of a formidable competitor for foreign capital as foreign direct investment would be channelled towards the reconstruction of eastern Europe. NAFTA was seen as the instrument through which Mexico would become more competitive by providing a more credible promise of economic stability that would appeal to the expectations of foreign investors. NAFTA is thus as much a treaty for trade as it is for foreign investment.

24 Neoclassical economics has produced some very fine theoretical models (Berkowitz and Kotowitz, 1982; Debrock 1985) demonstrating why in small, open economies with weak R&D investments *optimum* lifetime for patents should be short (not more than seven years). The reason is clear: in these cases, losses in welfare will not be compensated by protecting foreign patent-holders. Officials in charge of the Mexican Patent Office are not aware of developments in the literature, and this may somehow explain their arrogance.

25 They might have added that income concentration further reduced the domestic market and, together with oligopolistic industrial structures, favoured higher levels of idle capacity and higher unit costs.

26 Although productivity rose by 41 per cent from 1980 to 1992 the wages and benefits of Mexican manufacturing workers in 1992 were only 68 per cent of what they were in 1980. Average compensation totalled US$2.35 an hour in 1992, one-seventh of US earnings. This has been a cause of major concern for the US negotiators (OTA 1992) and was the major reason for extracting this promise from the Mexican side during the final stages of the negotiating process for NAFTA.

REFERENCES

Aboites, J. (1993) 'Cambio institucional y estratégias de ciencia y tecnología. La experiencia de Mexico en una década: 1909–1990', ORCYT-UNESCO, mimeo.

Berkowitz, M. K. and Kotowitz, Y. (1982) 'Patent policy in an open economy', *Canadian Journal of Economics*, vol. XV, no. 1, pp. 1–17.

Bueno, G. (1972) 'La estructura de la protección en Mexico', in Bela Balassa (ed.) *Estructura de la protección en paises en desarrollo*, Mexico, Centro de Estudios Monetarios Latinoamericanos (CEMLA). Also published in *Demografia y Economia*, vol. VI, no. 2, pp. 137–205.

Cardenas, E. (1987) *La industrialización mexicana durante la gran depresión*, Mexico, El Colegio de Mexico.

Casas, R. and Ponce, C. (1986) 'Institucionalización de la Politica Gubernamental de Ciencia y Tecnología, 1970–1976'. Taller de Investigación, Mexico, Instituto de Investigaciones Sociales, UNAM.

Comites (1975) 'Opiniones de los Comites de Ciencias Biológicas y de Ciencias Exactas del Plan Nacional de Ciencia y Tecnología', *Naturaleza*, vol. 6, no. 2.

CONACYT (1976) *Plan Nacional Indicativo de Ciencia y Tecnología*, Mexico, Consejo Nacional de Ciencia y Tecnologia.

CONACYT (1978) *Programa Nacional de Ciencia y Tecnología, 1978–1982*, Mexico, Consejo Nacional de Ciencia y Tecnología.

CONACYT (1990) *National Programme for Science and Technological Modernization, 1990–1994*, Mexico, Consejo Nacional de Ciencia y Tecnología.

CONACYT (1992) *Programa de Trabajo*, Mexico, Consejo Nacional de Ciencia y Tecnología.

Debrock, L. M. (1985) 'Market structure, innovation and optimal patent life', *The Journal of Law and Economics*, vol. XXVIII, no. 4, pp. 223–44.

Diario Oficial (1989) 'Acuerdo por el que se crea la Secretaria Ejecutiva del Consejo Consultivo de Ciencias como unidad de asesoria y apoyo tecnico del Ejecutivo Federal', Mexico, DF, 24 Jan.

Haro, G. (1967) 'El desarrollo de la ciencia en Mexico', *Espejo*, no. 2.

Izquierdo, R. (1973) 'El proteccionismo en Mexico', in *La economia mexicana (I. Analisis por sectores y distribucion)*, Mexico, Fondo de Cultura Económica.

Lozoya, X. (1973) 'Estado actual de la investigación científica en Mexico', *El Dia* (Suplemento dominical, numero 588), 30 September.

Lustig, N., Rio, F., Franco, O. and Martina, E. (1989) *Evolución del gasto público en ciencia y tecnología, 1980–1987*, Mexico, Academia de la Investigación Científica.

Maddox, J. (1993) 'Competition and the death of science', *Nature*, vol. 363, p. 667.

MIT (1989) *The Working Papers of the MIT Commission on Industrial Productivity*, vols I and II, Massachussetts Institute of Technology, The MIT Press.

Nadal Egea, A. (1977) *Instrumentos de politica científica y tecnológica*, Mexico, El Colegio de Mexico.

OTA (1992) *US–Mexico Trade: Pulling Together or Pulling Apart?*, Office of Technology Assessment, US Congress, ITE-545, Washington DC, US Government Printing Office.

Reynolds, C. W. (1973) *La economia mexicana. Su estructura y crecimiento en el siglo XX*, Mexico, Fondo de Cultura Económica.

Turckan, E. (1971) 'TUBITAK: a case for science policy in a developing country', University of Sussex, Science Policy Research Unit, mimeo.

Tyson, Laura d'Andrea (1992) *Who's Bashing Whom? Trade Conflict in High Technology Industries*, Washington, Institute for International Economics.

USTA (1976) *United States Trademark Association, Memorandum*, Office of the President to the Honorable Henry Kissinger, Secretary of State, from William Hedelund, President USTA. New York, 5 April 1976.

Part II

The politics of building science-based technology capability

Information technology policies in Latin America

5 A sectoral approach to changing technological behaviour

Weaknesses of Argentina's electronics and informatics policy

Hugo Jorge Nochteff[1]

The technological and industrial policy for electronics and information technology (EIP) implemented during the 1984–8 period, was the first – and until now the only – Argentine public policy explicitly focused on the development of a complex of mainly science-based and specialized supplier industries. It was also one of the few explicit sectoral industrial policies since 1976 and the only one which took the acquisition of technological knowledge as its central objective.

The singular nature of the EIP is even more noteworthy if the period of regressive restructuring and the inconsistencies of the policies in the period immediately prior to its formulation are taken into account. Its most specific feature is that EIP was an attempt to change not only the industrial structure but also the style of public policy itself. Whether it really gave special treatment to an industrial segment or not is debatable. In any case, for the few industries promoted and for those who benefited from the tariff policy, the EIP had some positive technological effects. In this respect, one can say that while as a sectoral policy EIP managed partially to reverse historical trends, as an attempt to change the general style of public policy and business behaviour it was more a failure than a success.

In this chapter I shall argue that the main, although not the only, reason for the EIP's limited results and eventual abandonment was its attempt to reform public and private styles of policy and behaviour via a sectoral approach and with limited institutional resources. In Section 1 the characteristics of the Argentine economy relevant for the subsequent analysis are briefly presented. The implications for science and technology policies of the financial orientation of firms and the regressive restructuring of the Argentine economy shape the context for an analysis of the informatics policy. In Section 2 the effects of regressive restructuring on the Argentine electronics complex are discussed and the main instruments of the 1984–8 electronics and informatics policy are analysed. Section 3 presents the discussion of institutional and political constraints affecting policy implementation. Policy results in relation to the technological behaviour of the companies that received promotion and the impacts of the policy on the electronic capital goods industry are analysed in Section 4. Finally, Section 5 contains a summary of the arguments presented and the conclusions of the chapter.

1 HISTORICAL CHARACTERISTICS OF THE ARGENTINE ECONOMY AND THEIR S&T POLICY IMPLICATIONS

The literature on science and technology (S&T) in Argentina agrees on some common points: that scientific and technological policies have never been a priority on the agenda of the state or society; that their formulation was belated and their implementation scanty; that there has been no stable connection between domestic supply and demand for technology (especially between private demand and public supply); that the technological behaviour of companies was basically adaptive and that, generally speaking, they did not seek to place themselves at the frontier of 'best practice', let alone change it. The scarcity or absence of technological and industrial policies in Argentina has been explained by the historical trends which have characterized both its economy and the behaviour of big business, by some features of import-substitution industrialization (ISI), and by the regressive industrial and economic restructuring which began in 1976.

The adaptation of Argentine economy to exogenous stimuli has characterized the economy's expansion periods throughout its history. This adaptation created non-transitory monopolies sheltered from competition by public policies and relatively independent of innovation, with the result that S&T concerns were almost absent from the agendas of government and big business.

In the two-sector macroeconomic model which characterized TNC-led import-substitution industrialization in Argentina, major innovations provided from abroad, albeit with a time lag, were mainly built into capital goods and inputs. Firms, geared exclusively to the domestic market, had to adapt their technology, scales and methods to the local situation. Thus the conditions were established for a low, adaptive and belated demand for technology. In this model, exchange rate and trade policies were of major importance in deciding: (a) the fate of firms, (b) the relative prices between tradables and non-tradables and within tradables, (c) the income distribution between agriculture and industry and between capital and labour. These policies also significantly influenced monetary and financial policies. Argentina's industrialization basically depended on these policies, and the promotion of scientific and technological development was relegated to a very secondary level. The two-sector model reinforced the effects of exogenously driven, adaptive economic behaviour on S&T policy (Nochteff, 1993).

There was a break in Argentina's history in 1976 which brought about profound political, economic and social changes. Of these changes, the two most relevant to this chapter are the growing economic importance of monopolies not based on innovation, and the regressive industrial and technological restructuring (Nochteff 1990, 1991b) which began in that year. In the following paragraphs I will analyse how the Argentine economy adapted to the favourable external financial conditions of the mid-1970s and early 1980s, resulting in increased foreign debt; how economic restructuring was brought about and what social consequences it produced. Distorted liberalization and strongly biased policies and subsidies were central to economic restructuring in Argentina. They have not only changed the behaviour of big business from industrial production towards financial activities, but also benefited a handful of locally-owned large economic groups and transnational corporations (TNCs) in a few industrial branches.

1.1 Consequences of Argentina's foreign indebtedness for changes in the behaviour of firms and S&T policies

Argentina's exploitation of the opportunity offered by the sharp easing of credit and the drop in international interest rates during the mid-1970s and early 1980s resulted in an increase in its foreign debt of about 450 per cent. Over the previous two decades the country had accumulated an external debt of US$7.9 billion, used to finance its growth. Within just four years, and without the Gross Domestic Product (GDP) growing in relation to its 1974 peak, Argentina's foreign debt increased by US$26 billion. The first period of serious indebtedness occurred between the middle of 1978 and the end of 1980, the majority of this indebtedness being private. Then, as creditors began to ask for better guarantees, public indebtedness grew at a faster rate and capital flight accelerated. Those companies who incurred foreign debt (mainly the large economic groups and some TNCs) brought in the funds, made financial profits, and then withdrew them, while the state put itself into debt by obtaining the foreign currency without which the reconversion from pesos to dollars would have been impossible. Finally, around 1982, the state took over private foreign debt and liquefied it by means of an exchange rate insurance system with interest rates very much lower than the currency devaluation rate (Basualdo 1987).

In contrast to the expansion of the second ISI, when adjustments followed shifts in world production and technological systems (although in a retarded and truncated way), the expansion phase which began in 1976 was a response to a short-term financial opportunity, ignoring all the technological and organizational changes that were leading to a third industrial revolution. Strictly speaking, it could be said that the Argentine economy geared itself away from (or uncoupled itself from) this revolution, at least as regards the composition of foreign trade and production structures (Nochteff 1990).[2]

The large economic groups and some TNCs, which incurred the bulk of the debt and then transferred it to the state, changed their organizational behaviour, with major consequences from an industrial and technological point of view. During the most dynamic ISI decade, the highest profits and the best growth opportunities had been in production activities in general and industry in particular. In the period 1976 to 1983, however, state subsidies and financial activities directly related to foreign debt (Basualdo 1987) or associated with portfolio movements (Damill and Fanelli 1988) provided the best profit opportunities. Under these conditions there was a shift in the focus of attention and priorities of firms away from industrial production – and the technological operations associated with it – to financial activities and lobbying (Nochteff 1991b).[3] The economy changed, altering the operational context of all firms, and not only of those that profited from the adjustment. This alone would have been enough for technological and industrial policies to be totally relegated. Trade and exchange rate policies and industrial restructuring strengthened and consolidated these effects.

1.2 Liberalization, regressive restructuring and the displacement of S&T in Argentina

In the early 1970s, at the closing of the last period of ISI, industrial growth, changes in industrial structure and increased productivity showed that progress had been made towards the formation of an industrial system in Argentina. One could then see that a basis had been provided for a possible soft landing from ISI to a more open and dynamic industrialization phase. This transition was, however, interrupted by the convergence of the world crisis and the Argentine economic crisis and the disruption of the political regime. In fact, during the military dictatorship which assumed power in 1976, the errors made in industrialization led to questioning of industrialization itself (Fajnzylber 1983). Since 1976, and especially since 1978, there has been a very sharp liberalization of the economy in which trade and exchange instruments have decided the course of Argentine industry. Trade liberalization has had two central aspects: the simultaneous appreciation of the peso and the elimination of trade barriers, and what I see as the double bias in the trade policies (Nochteff 1991a).

There was a sharp and quick appreciation of the currency between 1978 and 1981. Capital inflow decreased the exchange rate further, reducing industry's competitiveness in overseas markets and in comparison to imports, and causing relative price distortions in favour of non-tradables. The large economic groups and some of the TNCs diversified in two directions, outside and within the industrial sector, both consistent with the exploitation of financial opportunities, trade liberalization and the increases in the relative prices and profits of non-tradables. Outside industry, they went into services, especially financial services. Within industry, they concentrated on the more protected branches which involved lower commercial and technological risks. Among the latter were the less differentiated intermediate goods, especially those where competitiveness was based on natural resources or on externalities provided through public investment.

Argentine interest rates were kept very high (in comparison to international terms, i.e. measured by their dollar yield)[4] in order to attract foreign funds, in the initial phase of the indebtedness cycle, to avoid capital flight during the second phase of the same cycle, and because there was a divergence between domestic and international inflation. These interest rates, along with the shortage of local credit, generated two simultaneous processes. On the one hand, companies capable of incurring foreign debt to invest in the local financial market earned extraordinary financial profits. On the other hand, the less diversified industrial firms or those producing tradables paid high real interest rates.[5]

The trade liberalization policy was biased from two points of view. First, as the appreciation of the peso increased, the effective import exchange rate dropped less than the export exchange rate (CEPAL 1992), producing an anti-export bias. Second, tariff and non-tariff protection of the more monopolistic and scale-intensive activities were not reduced as much (and in some cases not at all) as protection for the science-based and specialized supplier sectors. This shows a second bias, this time between sectors. It is worth pointing out that this second bias explains, and at the same time is explained by, the shift in orientation of the large economic groups towards scale-intensive branches, especially less differentiated intermediate goods (from now on referred to as 'commodities').[6]

In one sense, the economic model of this period can be perceived as a typical cycle of short-term capital inflow (exchange rate appreciation, loss of competitiveness of the tradables, trade imbalance, increases in exchange rate risks and interest rates, and capital flight) followed by contraction and the loss of competitiveness in most of the tradables sectors and the industrial sectors in particular (Katz and Kosacoff 1989; Nochteff 1985, 1990).

In another sense, from the standpoint of domestic sales and exports, the biased liberalization increased the perversities of the two-sector model, producing a step backwards from the achievements of the 1964–74 period because it worsened the competitiveness of sectors which are more dependent on productive and educational organization than on natural comparative advantages. Furthermore, because the tariffs of the sectors which operated in competitive markets were reduced more than those of the scale-intensive sectors, which operated in oligopolistic markets and were protected by this market imperfection (Canitrot 1982), the former experienced greater pressure from foreign competition yet at the same time found themselves short of finance for reconversion.[7]

In addition to all these factors, industrial promotion subsidies for investment and operation reached levels hitherto unknown in Argentine history (FMI 1986). More than 80 per cent of these subsidies were given to a small segment of large Argentine economic groups[8] and were placed in scale-intensive industries whose competitiveness was based on natural resources and/or in capital-intensive externalities provided by the state (such as electricity or gas).

During the period 1976 to 1983, all the opportunities and public policies pointed to what has been described as the reversal of the infant industry argument (World Bank 1987). While the scale-intensive and some of the supplier-dominated, oligopolistic and natural-resources-related industries were growing or vertically integrating, the specialized suppliers and particularly the science-based industries were shrinking or disappearing (Katz and Kosacoff 1989; CEPAL 1987; Azpiazu, Basualdo and Nochteff 1988). This produced a significant step backwards in the activities that open the door to technical progress and innovation. Because of this effect, the economic restructuring initiated in Argentina in 1976 is said to be 'regressive'. The export performance of Argentine industry also deteriorated, as can be seen from data on manufacturing exports since 1974 (Azpiazu and Kosacoff 1987).

During the regressive restructuring period, all the factors and behaviours which had retarded the development of S&T activities and placed S&T and industrial policies low on the agendas of governments and big business were strengthened as never before in Argentine economic history. This radical displacement had some additional and interrelated causes. First, lobbying became much more important than during the ISI phase. It should be kept in mind that opportunities depended on government action (much of it ad hoc) such as the nationalization of debt, state guarantees for foreign debt, subsidies via industrial promotion, protection during trade liberalization, etc. Second, the economy and big business moved in the exact opposite direction to that of the last ISI period, shifting towards non-tradables and less differentiated industrial products manufactured in the scale-intensive and supplier-dominated sectors. In these two sectors, the main linking factors of technology demand – capital goods and inputs with higher technological content – were imported. With regards to big business, non-tradable activities became much more important than industry, and industries themselves were involved in activities that were protected, subsidized, dependent on natural comparative advantage, or which entailed low technological risk. Strictly speaking, big business moved away from dependency on imports. It is worth pointing out that process industries (steel, basic petro-chemicals, cellulose and paper, cement and aluminium) involved in the first stage of raw material processing, and protected as 'greenhouses' by public policies, are the least dependent of all industries on the existence of a national system of innovation. If the state provides some capital-intensive externalities (basically infrastructure and energy) and the company buys a turnkey plant, the majority of the competitiveness factors are assured. This does not occur with specialized supplier or science-based industries. This is why many technologically and industrially underdeveloped countries are competitive in petro-chemicals or iron and steel but cannot be competitive in machine tools or measurement and control electronics. The first two sectors can, as seen in Argentina, operate as enclaves; the latter cannot (Nochteff 1990).[9] In short, the protection enjoyed by the former sectors, and their weak connection with the rest of industry explains the designation of 'greenhouses/enclaves' (Nochteff 1991b). These sectoral characteristics, on the one hand, imply a very low direct technological demand on the national economic and technological system and, on the other hand, allow for the persistence of non-transitory monopolies which are not related to innovation.

Finally, the relocation of big business in greenhouses/enclaves produced changes in policy demands to keep or change comparative advantages. If the large domestic economic groups and TNCs had shifted towards tradables or importable goods, towards markets exposed more to competition and towards skill-intensive sectors in general, and towards specialized supplier and science-based sectors in particular, then, in order to maintain, improve or create comparative advantages, they would have required scientific and technological knowledge, increasingly better qualified human resources, industrial services and, generally speaking, all the components of the industrial fabric and the institutional system which supports competitiveness (Shapiro and Taylor 1990). In short, they would have required the formation of a national system of innovation. Conversely, because they shifted to non-tradables or non-importable[10] commodities and enclave sectors, these demands were insignificant or unnecessary. In this way regressive restructuring prolonged and aggravated the effects of the historical features of the Argentine economy and business behaviour with regards to the demand for and acceptance of technology policies.

2 PUBLIC POLICY IN ELECTRONICS AND INFORMATICS IN ARGENTINA (1984–8)[11]

The Electronics and Informatics Policy (EIP) formulated in 1984 consisted of non-tariff and tariff promotion schemes. The EIP non-tariff regulations were eliminated in 1988 with the adoption of new legislation for the Argentine industrial sector. Since then, there has been no special incentive for investment in the Argentine Electronics Complex (EC). The EIP tariff scheme was transformed by the reforms initiated in 1989, which accelerated the schedule of tariff reduction and established lower maximum tariff levels.

The EIP was formulated in an industrial and business environment defined by extraordinary financial profits, biased liberalization, debt crisis, flight of capital, subsidies and protection for commodities and enclaves; and this during a period of regressive restructuring. The previous section dealt with that period and its environment. The following paragraphs will discuss what happened with the EC during regressive restructuring, in order to detail some of the restrictions put on the EIP from a sectoral point of view.

2.1 The Electronics Complex (EC) and regressive restructuring

Between 1976 and 1984, in line with what happened to industry in general, the consumer electronics and telecommunications industries – the main local electronics industries[12] – were transformed into assembly plants for the domestic market, with imported designs, materials and capital goods. The parts and components industry practically disappeared, with production falling 91 per cent between 1974 and 1983. The number of engineers employed in the EC (Electronics Complex) decreased by 49 per cent between 1978 and 1983, and in 1983 the EC allocated only 0.45 per cent of its turnover to the salaries of R&D personnel.

Foreign trade in electronics confirms this involution and the dramatic fall in competitiveness of the EC. To begin with, between 1979 and 1986 Argentina's share in world imports of micro circuits fell by 67.5 per cent which, in the absence of a meaningful domestic production of components, indicates a significant decline in capacity to incorporate electronics-based technical progress into the Argentine industry. Furthermore, between 1970–2 and 1984–6, the ratio between the deficit and total volume of trade in electronic goods increased by 57 per cent, and the ratio between exports and imports fell from 0.35 to 0.23. If exports generated by IBM Argentina and their required imported inputs are excluded, the ratio fell from 0.27 to 0.04. This shows the virtual destruction of the EC's competitive ability. Virtually all exports are generated by the IBM factory which is involved in intrafirm exports of printers and other goods with a high electro-mechanical content, and operates under quite different conditions from other local EC firms. IBM's export factory is exceptional for several reasons: operations are based on stable and predictable demand (since purchases are intra-firm), foreign financing is at prime interest rates, requirements remain close to the 'best practice' frontier, and a temporary import scheme exists which includes exemption from import duties for inputs. Approximately 30 per cent of the added value of IBM exports is produced by local suppliers (usually small and medium-sized companies) which must keep their prices and technical capabilities very close to international levels. This is achieved primarily because IBM transfers some of the advantages mentioned above to local suppliers, so that they are operating in a quite different environment from the rest of the EC companies. In short, both IBM and its local suppliers partly operate as enclaves, although in this case they are virtuous enclaves.

2.2 Formulation of the Electronics and Informatics Policy (EIP)

In 1984 the Executive created the National Commission for Informatics (CNI) to make policy recommendations to the President for the entire electronics complex (EC). The CNI consisted of representatives of eight public agencies whose areas of expertise covered the whole institutional spectrum required to implement a technological and industrial policy. The CNI recommendations were presented in October 1984.

With regard to the supply of electronic equipment, the CNI recommended[13] support for local production via the promotion of selected industries. The target was to be those goods whose manufacture, due to the level of technological complexity and the investment required, was accessible to local industry via associations with foreign companies. As far as the commission was concerned, technological knowledge and human resources development had to be the prime objectives, in order to create a competitive industrial complex, to speed up the diffusion of new technology, and improve its positive effect on economic competitiveness in Argentina. The supporting documents to the recommendations stated that priority must be given to technology because the objective was not only to learn to produce but also to learn what to buy. Promotional incentives were to be offered by means of competitive bidding, open only to companies with predominantly domestic shareholders, and which would undertake to meet production performance, technological and price targets.

As regards demand, the CNI recommended that state purchases should give preference to the promoted companies and products. It also stated that the most efficient way of incorporating hardware was by adopting policies for standardizing software, for promoting system compatibility, and for increasing the use of distributed processing systems. To implement its recommended strategy, the CNI proposed the creation of a National Commission for Informatics and Telecommunications (CONITE), comprising those agencies that had formed the CNI, to co-ordinate the policies and operations of the state agencies.

In the following paragraphs the EIP will be analysed in its three major components: the tariff system, the industrial promotion scheme and state purchase.

The tariff system

The purpose of the tariff scheme was twofold: (a) to reverse the situation produced by biased liberalization by increasing protection rates and (b) at the same time, to avoid distortions that were particular to the ISI stage. To this end, the non-tariff barriers imposed since 1983 were replaced by a system of tariffs with decreasing levels according to three criteria:

1 time (starting with a maximum tariff of 100 per cent in 1986, with annual reductions to 50 per cent in 1991);
2 stage of production (maximum tariffs for finished products and sub-assemblies, gradually reducing for components); and
3 production opportunities (higher tariffs for products for which local production was more feasible).

The features that distinguish this tariff system from other systems applied throughout Argentina's history are as follows:

1 As a component of the sectoral industrial policy, the idea was to link it with other components of the policy (basically with the sector promotion scheme).
2 It set equal tariffs for capital and consumer goods, and for finished products and semi-knocked-down sub-assemblies.[14]
3 It restricted the use of effective protective rates to a narrower range of products and reduced the size of the differences between classes of nominal tariffs which had distorted price/performance ratios.

4 The effective protective rate tended to be more closely aligned to the corresponding added value (until 1984 the most protected segment of the EC had a zero or negative added value at free trade prices).
5 Tariff-fixing was based on public and objective criteria which made the protection system less susceptible to private or political pressures and contradictory administrative decisions.
6 It gave local producers only a temporary opportunity to consolidate their competitive position (because tariffs were decreasing).

In point of fact, these were not monopolistic conditions but they encouraged companies by helping them, during a given period, to cope with market imperfections (technology suitability and economies of scale) and thus to consolidate favourable, competitive positions. The most significant difference, as compared to the previous systems, is that until 1984 effective protection was biased in favour of activities with low added value and comparative disadvantages (static and dynamic), while the tariff system implemented in 1985 shifted the effective rate to activities with higher added value and comparative advantages (Nochteff 1993).[15]

The industrial promotion scheme

The EIP (Electronics and Informatics Policy) promotion scheme was also radically different from others that had been applied and were still being applied in Argentina. It was directed towards a partly science-based and partly specialized suppliers' industry, generating dynamic comparative advantages. Also the main objective of the policy was the acquisition of technical knowledge instead of the granting of subsidies to mature, declining, low-added-value industries or those based on exploiting natural resources which incorporated technology mainly through imported inputs and capital goods. The previous statement does not necessarily apply to the segment of lower-cost microcomputers. Its inclusion among the products selected for the informatics industry promotion scheme was probably the EIP's greatest technical error. In any case, until 1984 microcomputer production required significantly more skilled labour, engineering, R&D and specialist suppliers than the 'commodities', iron and steel, petro-chemicals, cellulose and oils which had been promoted (much more intensely) in Argentina since the mid-1970s.

Promotion occurred by means of open, transparent, competitive bidding in which the evaluation criteria were public and objective, so that benefit grants were dependent on project quality rather than lobbying ability. It should be stressed that, although this mechanism had been contemplated in the existing legislation, it had never been applied before. The ratio between subsidies and private investment was much lower than that of the other

Argentine promotion schemes, due to the type of benefits granted, the fact that they were decreasing, and that they had to comply with performance, R&D, price and other requirements.

The EIP was formulated for the whole of the electronics complex. It was intended to form an articulated technological and industrial system, taking into account externalities and the internal and external linkages of the complex. In this way an attempt was made to overcome both the industrial disarticulation provoked by regressive restructuring and the creation of non-transitory monopolies which were not related to innovation and were weakly linked to the industrial fabric. Although the promotion system was not implemented in the whole of the electronics complex, the informatics scheme promoted the articulation between segments and projects, such as links between producers of different goods or between these producers and system integrators. By promoting these links, the scheme aimed at forming economies of scale and scope and an industrial complex. This, if implemented for the other segments of the complex, would have reversed the changes in the Argentine EC during the 1976–83 period, which have been described as a conversion from an industrial system into a set of enclaves.

Temporary support was granted to local firms to consolidate their market positions. The scheme tried to encourage competition, as can be seen not only from the above-mentioned temporary promotion and protection, but also from a reduction in market entry barriers and an increase in the number of suppliers in certain oligopolistic markets (such as the major state computer centres). This last issue (together with the inclusion of technology) was one of the reasons why an attempt was made to encourage the establishment of domestic industrial companies and to attract TNCs that had not managed to penetrate the local market (or which had an insignificant presence). A specific segment was reserved for small companies that had been practically excluded from previous promotion schemes. These companies could participate in the informatics scheme if they met the policy's established conditions which, at the same time, produced a method of 'picking winners' and stimulated competition.

Firms were given a set of requirements in terms of performance, prices, integration of products and processes, investment in R&D activities, use of trademarks and marketing structures. These requirements were very much more demanding than those of other previous and ongoing promotion schemes. These requirements encouraged a production path quite different from those that had been predominant in the Argentine EC. In fact, they made the transitory quasi-monopolistic conditions which were provided through tariffs and promotion dependent (as and when the advantages were reduced) on the companies' competitiveness and capacity for innovation.

The handling of foreign investment and the relationship between TNCs and local firms were also quite different from what was usual in Argentina. In the first place, promotion was reserved for companies with a majority of domestic shareholders. Second, the requirements to be met in order to earn promotion were especially demanding with regard to the relationship between domestic and foreign partners in matters such as technology transfer within partnerships, the use of licences and trademarks, the prices of licensed products, and the marketing system. The majority of the projects were obliged to develop, after a given period, a technologically advanced version of products which had been licensed and to launch them on the market under their own trademark. These requirements show that the concession of quasi-monopolistic advantages was meant to be temporary and related to the capacity for innovation.

Finally, this was the only scheme specifically to promote the production of capital goods, in contrast to the earlier environment which had systematically discouraged it.

State purchases

The EIP attempted to increase the transparency of state procurement, reduce entry barriers for state suppliers, and encourage a convergence between the public sector demand and the supply from the firms promoted by the policy. To do this it was necessary to create standardization mechanisms, to discourage purchasing 'packages' or purchasing on a turnkey basis, and to effectively enforce the 'buy domestic' scheme. All these conditions were to be met during the phase of defining state demand, when needs would be determined and alternative technological solutions analysed. The provision of such information to the local industries would facilitate their access to the public market and increase the number of state suppliers.

The strategy chosen to encourage these changes in the state's purchasing methods was to establish a negotiating process involving the Secretary for Industry and Foreign Commerce and the Undersecretary of Informatics and Development, on the one side, and state-owned companies, on the other.

3 CONSTRAINTS AND OBSTACLES IN THE IMPLEMENTATION OF THE ELECTRONICS AND INFORMATICS POLICY

The EIP's implementation faced restrictions which completely altered its development and results. In my view these restrictions are ultimately consistent with:

1 Argentina's past development under an exogenously driven adaptive economy, in which innovation had only a secondary, if any, importance;
2 the conduct of big business, especially the growing trend to strengthen the formation of non-transitory monopolies;
3 the pattern of state priorities, and the policies and resources related to this behaviour; and
4 the restructuring which began in the mid 1970s.

Other major constraints, such as macroeconomic instability or lack of political support – especially the latter – are to a certain extent dependent on the four factors listed above.

Consistent with all this, research has shown that after the first stage of formulation, the EIP was not a priority on the state agenda (Azpiazu, Basualdo and Nochteff 1988, 1989, 1990a, 1990b, 1991a, 1991b; Nochteff 1985, 1990, 1993). Interest in it and commitment to it by the major economic agents was very low, and local technological and industrial regression restricted firms' development.

The EIP is commonly seen as a policy to promote the data-processing equipment industry ('informatics') because this was the only EC industry which actually received promotional benefits. However, a different picture emerges from an analysis of EIP's formulation and implementation, showing that its explicit and implicit objectives were aimed at the entire EC, with the exception of assembly activities for the domestic consumer goods market. First, the policy's scope is rooted in the CNI recommendations which were directed at the whole of the EC. Second, the promotion scheme included inviting bids, in successive rounds, for promotional benefits for data-processing equipment, telecommunications, measurement and control electronics, etc. It happened that only the first bid became a reality. Third, the tariff policy encompassed all electronic goods, yet it could not change the effective rate of the *maquiladora* of consumer goods since this sector had its own import duties exemption regime. The tariff policy was in force for four years, and research on the EC's productive and technological performance at that time shows that, despite the amendments and problems it underwent, the policy had a strong positive impact on the whole of the electronic capital goods industry and not only on data-processing equipment (Nochteff 1994). Actually, its positive effects were stronger in the segment of measurement and control electronics (industrial and medical) than on data-processing equipment. The analysis of the latter sector, the only one to which several promotion instruments were applied, enables us to appreciate more clearly the restrictions put on the EIP.

3.1 Institutional and political restrictions

In so far as it involved the use of different instruments and negotiations with very diverse actors, the EIP required several state agencies to act consistently and with similar criteria. However, the co-ordination body recommended by the CNI, the CONITE, was never formed. Although the CNI's recommendations had been produced at the request of the President, most were not implemented, with no reasons given. Furthermore, the proposal put forward by the Chamber of Deputies to the Senate was never discussed and the Deputies did not insist on the issue. Actually, some legislators of the government party were enthusiastic about the policy in its early stages, while the opposition criticized it for being low profile, and demanded a more interventionist, nationalistic and 'Brazilian-style' strategy. But afterwards both parties lost interest or did not support what they had at first considered to be a low-profile alternative. The Secretary of Industry and Foreign Commerce and the Undersecretary of Informatics and Development continued to push the policy, which finally became the responsibility of the former.

The fact that CONITE could not be formed and that none of the government's higher echelons took on the co-ordination of the different agencies involved is an indication of the low priority given to this policy, and is one of the factors explaining the serious problems in implementing it. In effect, the Secretary of Industry and Foreign Commerce had to negotiate with many government agencies that considered decisions affecting the electronics and informatics industries to be of secondary importance. This situation was the immediate cause of the lack of co-ordination and of the contradictions between the various legal instruments and between the various government actions. However, what ultimately happened confirms the persistence of the Argentine public policy style and of institutional behaviour prior to the EIP. This was a style and behaviour consistent with an economy featuring non-transitory monopolies and enclaves, which did not require an institutional system to create links between the technological and industrial objectives, the S&T agencies and the industrial policy, the state-purchasing regulations, the industrial promotion legislation and the credit system.

3.2 Policy inconsistencies and contradictions

The problems of the tariff policy

The original tariff policy underwent major alterations, the effects of which were compounded by a lack of co-ordination with the industrial promotion policy. The main effect of these alterations was to reduce the actual protection provided.

First, tariff reduction was incomparably faster and more far-reaching than had been forecast, because of general rather than sectoral decisions: the maximum tariff for electronic capital goods, which was to be reduced gradually to a minimum of 50 per cent in 1991, was less than half of that (24 per cent) by August 1990. On the other hand, due to delays in the promotion scheme, most of the firms selected from the bidding began to receive promotion benefits only between January 1987 and January 1989. Consequently, none of the firms selected received promotional benefits while the initial maximum tariff was in force, and the majority received them when the tariff schedule was already between its second and third years. Moreover, as will be seen below, the rest of the electronic capital goods industry (measurement and control, telecommunications, etc.) never received any promotional benefits. In short, a consistent combination of the tariff and promotion policies was never achieved.

Second, the first formulation of the tariff policy enabled firms to import certain components and materials at reduced tariffs as long as the local production of those inputs could not achieve an adequate price–performance ratio. This was co-ordinated with the promotion system, which encouraged the local production of certain of these inputs. However, the reduced tariffs were applied very late, affecting the expectations of the market and the firms, and especially prejudicing those firms which were already manufacturing computer products. Owing to the legal interpretation of the agencies answerable to the Treasury, the reduced tariffs established by the Secretary of Industry and Foreign Commerce were removed almost immediately after their grant had been announced, under another legal form, between one and three years later than planned. As a consequence, computer manufacturing firms were operating for at least a year with much higher input tariffs (over 50 per cent) than those forecast (between 15 per cent and 30 per cent).

Third, following the pattern of events for the whole industrial sector during the period 1976 to 1983, non-tariff protection in the EC inverted the infant industry argument since it benefited mature and/or assembly industries. On the one hand, while non-tariff protection was removed for most of the EC in 1985, 47 per cent of all other goods in 1986 and 35 per cent in 1988 were still receiving such protection. As regards electronic goods, the only ones that continued to enjoy non-tariff protection were consumer goods which, as noted above, were produced by assembling industry (*maquiladoras*) and destined for the domestic market. On the other hand, specific duties were applied from the beginning of 1988 which gave high levels of protection to these assembly activities and many other industries (areas of pharmaceuticals, automobile, basic petro-chemicals and plastics) which had maintained informal market reserves for decades. When non-tariff restrictions were removed, it was discovered that the *ad valorem* tariffs on an enormous number of products were insufficient, mostly due to the inefficiency of the customs system. Therefore other protective mechanisms, such as specific duties, were implemented. Specific duties were imposed for most of the products that had been removed from non-tariff protection in 1988. Nearly 150 specific tariffs were brought into effect in a single month that year, *but not for the electronic products promoted by the EIP.*

To sum up, contrary to what might appear from a superficial analysis of *ad valorem* tariffs, the electronic capital goods industry – including informatics – was one of the least protected industries, and when it was protected this was only for a short time, despite the fact that it was recognized as an infant industry. Moreover, the tariff policy continued to protect monopolies unrelated to innovation.

In spite of the above, the tariff policy for all electronic capital goods, parts and components, was the first to come into force and the last to be abandoned. This confirms that in Argentina (as in many other countries) trade policy was always the greatest promotional instrument in the various stages of industrialization. Trade policy-makers could design, implement and keep policies in force longer since they demanded fewer institutional resources and innovation. The civil service saw solutions to problems through the instruments at their disposal, and their options were limited by the tools they had at hand. The best way to promote certain technological skills, given limitations in the instruments available, was to shift the maximum levels of effective protection rates.

The industrial promotion scheme

Changes in the general industrial promotion legislation, necessary to achieve consistency with EIP, was delayed by political and economic pressure, and began three years later than had been planned when the EIP was drafted. The reform of general promotional legislation and some of the specific schemes were delayed by the resistance of the political opposition parties, a few provincial governments and the bulk of business. The question was so difficult that in 1992, nearly a decade later, many of the subsidies had still not been removed. This situation led to several constraints. First, some of the objectives could not be implemented (such as locating plants near R&D centres) and the application of some instruments (such as exemption from or reduction of input tariffs) was seriously altered and delayed. Second, the delay was one of the reasons for the lack of co-ordination between the promotional benefits and the tariff policy. Third, it contributed to the failure to call for bids for benefits to be granted to other electronic capital goods industries.

Furthermore, exemption from Value Added Tax (VAT), one of the main benefits offered by the promotional system in its first version, was never granted to firms who succeeded in winning bids. This was due to changes in the Treasury's criteria with regard to the new promotional systems. Paradoxically, competing companies that had not entered the informatics promotion bid, or that had not qualified for it, received promotion with VAT exemption granted under other existing systems, with no obligation to comply with requirements of industrial integration, R&D and improvements in the product and production process such as was required of the winners in the informatics scheme. Although the 'winners' of the informatics promotion bid received other benefits, such as partial exemption from input tariffs (with the aforementioned delays), the factors above altered the incentives mix and consequently the project profiles and the policy results. The unfavourable coexistence of the EIP with other promotional schemes confirms the lack of coherent technological and industrial policies, instead of which non-innovation-related, monopolistic advantages have been granted.

The public sector demand

The Secretary for Industry and Foreign Commerce and the Undersecretary of Informatics and Development, working with the existing 'buy domestic' Act (which in practice did not cover technological questions) could not change either the direction or the terms of state purchasing. Public sector demand was oriented towards turnkey systems and the importation of technology and capital goods, especially R&D-intensive goods, which were usually provided by a handful of big supplier companies. This orientation was in fact reinforced, particularly among public financial institutions whose modernization process made them the most dynamic informatics market in those years. It is important to emphasize that these features of state demand were associated with a growing concentration of state suppliers and a deterioration in the technical and managerial capacity of state-owned companies and agencies during the period of regressive restructuring and consolidation of non-transitory monopolies.

This, together with the unexpected fall in state investment, was one of the major factors adversely affecting the EIP. Implicit in the informatics policy was the assumption that 1985 would be the turning point in the economic cycle. This assumption was shared by the vast majority of analysts and decision-takers, on the basis of the initial success of the so-called Austral Plan. However, in 1987 the recession deepened and public investment again began to fall. Originally, it was planned that demand for equipment would essentially come from official financial institutions, but the fall in public investment and especially the impossibility of reorienting state purchasing reversed these projections, drastically affecting the sales levels of promoted companies. Demand from the state financial sector, crucial for the development of some of the main projects, was less than 30 per cent of the whole demand from the financial system in the first year (1987), and only 4 per cent in the second operational year of those projects (1988). What happened with public sector demand illustrates the weakness of the explicit policies and of the political influence of civil servants in support of policy reform *vis-à-vis* traditional patterns of priorities and behaviour.

3.3 The weakness of the Electronics Complex

During restructuring, the conversion of the whole of the consumer electronics industry and the majority of the telecommunications industry into merely assembling activities caused the specialized suppliers' network virtually to disappear. In addition, all but a very few specialized suppliers who continued operating in these industries cut their investments, in general and in technological modernization.

When the EIP was drafted, policy-makers sought to remedy this weakness in two ways. First, a programme was adopted to increase gradually the domestic input content in equipment. Second, the programme was to run parallel to an increasing demand for domestic input in the consumer and telecommunications industries in order to achieve economies of scale and scope in the production of mechanical parts, electronic components and sub-assemblies. As these finished product industries accounted for over 80 per cent of the input demand of the EC, it was felt that a moderate increase in the purchase of inputs in the domestic market would be sufficient to drive rapid growth and upgrade the supplier base. This would fulfil the needs of promoted companies as and when they had to comply with programme requirements to increase the local content of equipment.

To achieve this it was necessary to change the Tierra del Fuego promotion scheme, in force since 1972, which created an exemption from import tariffs on intermediate and capital goods for companies located in that island. These incentives did not attract consumer electronics companies until 1979. In 1983, however, 89 per cent of all Argentine consumer electronics companies were located in Tierra del Fuego (Azpiazu, Basualdo and Nochteff 1988: 88). In addition, the state-owned telecommunication company (ENTEL) would have to be more clearly oriented towards increasing local content requirements in their purchase contracts. In 1984, both changes were considered imminent.

The Tierra del Fuego scheme had been severely criticized. The main opponents of the system were the Secretary of Industry and Foreign Commerce, some legislators from the two major parties, and electronics firms with operations outside the promoted area. Besides, the legal time limits for tariff exemptions had almost expired and discussions on amendments and possible renewal were necessary. Finally, local content requirements could be increased by means of an executive decree without changing the law. Despite the criticism, tariff exemptions on inputs were extended, apparently for geopolitical and demographic reasons. It is worth pointing out that: (a) the effects for the industry of the incentives behind the Tierra del Fuego promotion scheme were not taken into account; (b) a few of the incentives whose fiscal cost was lower (although more explicit) than the cost of input tariff exemptions were removed; (c) both benefits and exemptions that were most prejudicial to economies of scale and specialization in the EC were retained. This definitely points to an implicit pattern of priorities excluding sectoral and general industrial and technological issues.

It was also expected that the local content of ENTEL's purchases would significantly increase when its principal contracts were renegotiated. An increase of local content had been the explicit objective of the Secretary of Communications (SECOM, the agency to which ENTEL was responsible), and recommendations of the CNI and SECOM on the telecommunications industry had been rubber-stamped by two executive decrees. None the less, contracts were renegotiated without including obligations to buy local inputs. Instead priority was given to the speed at which lines could be installed (Herrera 1989). Again, technological and industrial issues were put aside.

The inability to change the regulation of assembling activities prevented any convergence between the supply and demand of domestic inputs for EC industries, which was both a policy objective and a condition for its success. Moreover the non-electronic, specialized supplier base was undermined. Consequently, there was only limited local content in the informatics industry. Firms that did attempt to increase local content faced sharp rises in costs.

3.4 The behaviour of the local major economic agents

As was seen in Section 1, before the advent of EIP the large Argentine economic groups and some TNCs acting in various economic fields had diversified and shifted towards non-tradables and monopolies with low technological risks. Within the EC, the trends in orientation of these groups were: (a) shifting into the mere assembly of final goods destined for the domestic market; (b) emphasis on importing operations; and (c) establishment of partnerships between TNCs and large Argentine economic groups, the latter contributing their lobbying ability and/or civil engineering capability. The EIP attempted to change this behaviour by encouraging Argentina's large economic groups to become majority partners in industrial projects, going beyond mere assembly. Their monopolistic advantages in such projects were temporary and dependent on a capacity both to absorb technology and to innovate.

Traditional behaviour did not change. First, the number of Argentina's large economic groups who participated in the promotion scheme was very low, even if one counts those that were involved in electronics and informatics only as importers and/or service providers. Second, the large economic groups who did participate adopted a passive attitude and had a low commitment to this system. On a few matters, such as state purchasing or the joining of several EC sectors, their position was ambivalent. This passivity contrasted with their active and public involvement with other industrial promotion matters, state purchases, nationalization of private foreign debt, or subsidies to the financial system. In these matters they were decisive in defining and implementing many public policies, despite the fact that all of these policies entailed levels of complexity, decision-making and political resources incomparably higher than those of the EIP. During the regressive restructuring period, these public policies were also decisive for activities involving higher profits (Azpiazu and Basualdo 1989).

The large Argentine economic groups participating in the informatics projects, and those which had already gone into other EC industries, did not attempt to build articulated production and technology strategies in electronics, despite their involvement with supplying telecommunications equipment and with the consumer electronics industry. This involvement would have enabled them to push ahead with restructuring the supplier base which indirectly relied on the informatics projects.

As regards TNCs, the investment strategy of those operating in other segments of the Argentine industrial market in that period did not include electronics (Basualdo and Fuchs 1989). There is no evidence of their opposition to EIP or their support for it. On the other hand, the TNCs who did have investments in the Argentine EC, but not in the segment of data-processing equipment (i.e. Siemens, NEC, Phillips), and which also did not plan to enter the informatics segment, feared the policy spreading to telecommunications and consumer electronics. There is no evidence that their 'disapproval' of EIP had turned into active opposition to the policy, despite the fact that, benefiting directly from public contracts in telecommunication or from the Tierra del Fuego scheme, they might have felt threatened by the changes EIP would imply for these promotion schemes. A more supportive attitude from TNCs came from those who saw in the EIP the support they needed to enter the local informatics market, which was thoroughly dominated by IBM. These TNCs eventually entered the industrial promotion scheme.

In 1985, IBM Argentina was practically the only company in the country manufacturing informatics goods. This company was the main producer in the local EC and generated over 90 per cent of the production in the computer and office machinery sector. It was also Argentina's biggest industrial exporter and generated between 85 and 90 per cent of electronics exports. IBM Argentina also virtually controlled the supply of data processing equipment to the state, using imported equipment and systems (except for printers). Briefly, IBM's strong and permanent opposition to the EIP was very relevant, because of the pressure the company could exert not only because of its size on an international scale but also because of its domestic importance as a producer, exporter and state supplier. As a producer and exporter it did not need an electronics policy, but, as an importer and state supplier, it could be affected by one. Nevertheless, it was with an IBM licence that Argentine Microsistemas entered the industrial promotion scheme. This might indicate that, notwithstanding its disapproval of the EIP, IBM decided not to risk being left outside of it. Other TNCs such as Unysis, Bull and Stride, while not being strongly interested in the policy did not oppose it. The EIP could give them an edge to compete with IBM in the local market. Their interest was not stimulated particularly by the fact that both the local conglomerates and the state soon lost interest in the policy. Nevertheless, Unysis and CNL Bull entered the scheme and Unysis was among the last big companies to leave it.

The attitude of the large Argentine economic groups was perhaps the single most important factor in deciding the fate of the EIP. The inability to overcome their uncommitted attitude demonstrates the weakness of the group of civil servants who promoted the EIP. The strong reformist nature of the policy with respect to historical Argentine public and private behaviour and the industrial structure required considerably more than the institutional and political resources provided by the co-operation of two government agencies (or three, if we include SECOM), the technicians of two technological centres, and a group of small or at best medium-sized innovative industrialists. Although it might appear that the attitude of the civil servants was naïve or that they overestimated the support of the President and some legislators, an analysis of the vicissitudes of the EIP and the interviews[16] held show that they were also confident that growth opportunities in electronics and informatics would attract the large Argentine economic groups.

The political wager seemed reasonable when some of the large economic groups entered the informatics promotion bid. However, the interviews also show that the interest of large Argentine economic groups was very low and short-lived. Some of these groups appointed officials or designated areas in each conglomerate to follow EIP, thereby making an initial undertaking to participate in informatics. They were interested in the dynamism of the sector, the possibility of winning state purchase contracts – basically from the financial institutions which were re-equipping – and in consolidating positions to enable them to cross entry barriers if the EIP was successful. But these major companies were very diversified: not only did their main sources of profit lie in other activities, but they had also been accustomed to modalities of profit-making virtually the reverse of those stimulated by the EIP. This contradiction was particularly clear in relation to the large economic groups who had smaller assembling companies acting in the EC. The size of the groups in question, their diversification levels, and the areas in which they earned the bulk of their profit (and almost certainly the highest rates of return) suggest that their participation in the EIP was, so to speak, a small wager: if the policy was successful in itself they would add another activity to the conglomerate; if it was not they would lose very little (only two of them actually made an investment, significant in terms of the EIP, but not in terms of the conglomerate itself). In short, the involvement of large Argentine economic groups did not include the contribution of their political and economic weight but only, at best, the participation of a division or a small company from each of them. The officials who pushed the policy never managed to get the resources or the minimum private support and links for such an ambitious reform (at least on a sectoral scale).

It was this lack of support from the major Argentine economic groups, together with the extremely weak support from the political system (government and opposition), which explains the abandonment of the EIP. It is very probable that strong support from the political system would have strengthened the interest of the large economic groups. It is more probable (perhaps certain) that strong interest from those groups would have attracted both the attention and the support of the political system and some TNCs. When the large Argentine economic groups showed little interest and low commitment, the TNCs involved in the scheme (other than IBM) began to lose their initial enthusiasm. At the same time, IBM's opposition became more decisive, not so much because of the company's sheer strength but because there were no powerful actors on the other side of the arena. Furthermore, the conflict with IBM also depended on the EIP's progress, because if the promotion policy, as envisaged in 1984, could have been extended to other EC sectors (measurement and control, telecommunications, etc.) it would have been easier to accept some of IBM positions in informatics, or at least to achieve a satisfactory compromise. Conversely, the narrower the scope of the promotion policy, the higher the cost of any trade-off between the EIP and IBM.

3.5 Gradual abandonment of the EIP

These constraints increasingly limited the EIP. In the informatics sector, companies steadily withdrew from the promotion scheme. Some did not present proposals for the final benefit awards. Others either did not make use of these benefits or deferred their investments. Still others finally gave up the promotion and closed down their plants. The other sector schemes (telecommunications, industrial electronics, etc.) were also eventually abandoned. This was due to three factors:

1 In some sectors (e.g. telecommunications) larger companies did not need or want the scheme since state purchases under contracts which did not demand performance were assured.
2 The few entrepreneurs who supported the EIP withdrew their support in light of the obvious political weakness of the administering officials.
3 There were major changes in the team of officials.

Some of these changes were due to pressure from firms who felt their oligopolistic positions being challenged or their performance requirements being increased. Other changes were due to government priorities which did not coincide with those of the EIP. A significant example was the change made in the team of officials at the Secretariat of Communications. These changes came about for two reasons:

1 There were confrontations between these officials and the two major state suppliers, who opposed an increase in the local content clause of their supply contracts proposed by state officials as a way of upgrading the supplier network (a necessary condition for EIP success).

2 There was the decision of the Executive to prioritize the rapid installation of lines over and above any industrial or technological considerations, which had been the criteria applied by state officials in the telecommunications sector since 1976.

The EIP tariff policy was amended in 1989 and finally abandoned as part of a new liberalizing orientation which led to the general reduction of tariffs. None of the other EIP aspects (such as human resources or R&D institutional development), even more alien to the Argentine traditional style of state intervention, were implemented.

4 EFFECTS OF THE ELECTRONICS AND INFORMATICS POLICY, 1884–8

Having analysed the constraints that seriously limited and finally brought about the abandonment of the EIP, I will briefly discuss its impacts on the data-processing equipment industry (informatics), the only one which actually received promotional benefits, and on the electronic capital goods industry as a whole (computers, telecommunications, medical and industrial electronics) which benefited from the EIP tariff policy.

4.1 The effect of the EIP on the informatics industry: projects and results

In June 1985 the Secretary for Industry and Foreign Commerce evaluated company proposals to produce informatics goods under the EIP specific promotion scheme. The number of proposals submitted (51) was more than double the awards envisaged (25). Although the preliminary awards were made known in October 1985, the operations of the first project only started in January 1987, at the same time as the final award of the benefits. The delay in the award, the changes in promotion incentives and in the tariff system, and the problems in changing the direction of state demand discouraged private sector interest in the scheme. For these reasons, twelve of the selected twenty-five firms withdrew before the final award.

The thirteen firms who finally received benefits were of two types. On the one hand, there were five large economic groups, three of which were associated with TNCs (Stride, Unisys and Bull) while a fourth had an IBM licence to produce banking products. On the other hand, there were eight smaller local firms involved in electronics, some of which had concluded technology transfer agreements with foreign companies (such as Star Micronics, TEAC, Olivetti and Spectraphysics). It should be emphasized that only one of the large economic groups had actual production and technological experience in electronics, in this case acquired through the takeover of one innovative and high-performance medium-sized firm (the one which held the IBM licence). Two others had some experience in assembling and sub-assembling activities. All the smaller firms were very experienced in electronics, and were some of the best performing firms in the EC as producers, systems integrators and, in some cases, importers.

On the basis of the proposals made by these thirteen firms it is possible to estimate, in general terms, the investment, sales, exports and employment figures forecast for the end of 1986 (when the awards were given). The investment necessary for an industrial plant to start operating was in the region of US$44 million, while a total of US$93.3 million would have to be invested during the first six years of operation. Estimated sales for the first six years were US$1,400 million (approximately 40 per cent of the projected market of data-processing equipment). Firms were required to export 6 per cent of production. In the first six-year period, maximum employment was estimated at 3,300 jobs. Professionals and technicians (more than 300) to be employed in R&D and quality control represented nearly one-tenth (9.2 per cent) of the total estimated employment in the promoted plants, which meant more than a 50 per cent increase in demand for professionals and technicians in those areas in the whole of the EC in Argentina.

In relation to technological activities, the difference between the electronics industry which this policy was trying to promote and the existing one was clear. The estimated employment figure in R&D and quality control (9.2 per cent) for the promoted electronics industries was nearly three times higher than the actual figure for the whole of the local EC in 1984. This difference was even clearer in terms of R&D expenditure, which in the promoted projects was allocated nearly US$60 million over the six-year period. This was equivalent to 4.2 per cent of total actual sales, more than five times the actual R&D expenditure rate of the local EC at that time.

Because of constraints on the EIP, and its abandonment in 1989–90 not much can be done to assess its eventual impact on the informatics industry. In fact, in March 1990, only three of the projects were still running within the promotion regime. Approximately half of the remaining ones had never managed to get fully under way and the other half had totally or partially halted production because of the problems described above.

However, if we examine some information about the three firms that were still operating in 1990 – Microsistemas, Autorede and ICSA – we see some interesting features. Microsistemas had a significant previous history in informatics as an independent company, before being acquired by a large economic group. This company automated virtually the whole of its manufacturing plant between 1986 and 1988 and employed 215 people, of which 31 per cent were professionals and 34 per cent technicians (65 per cent of the professionals were employed in production, quality control and R&D). In 1988 its sales reached US$22 million, five times higher than in 1986, but 75 per cent lower than forecast in the promoted project. Some 80 per cent of sales were related to products for the banking market (terminals, controllers and peripherals) and 20 per cent to general purpose products (microcomputers based on Intel 80286 and 80386), thus confirming the importance of financial system demand. However, state banking demand was very low, contrary to what had been envisaged by policy-makers, but consistent with the difficulties which have been described in attempting to change the public sector's traditional buying behaviour. In 1988, exports reached US$2.5 million, doubling the forecast exports/sales ratio.

Towards mid-1989, Autorede fulfilled the majority of local content and R&D requirements. Its sales in that year were US$1.5 million, and 35 per cent of all employees (40 people) were professionals or technicians. As in the case of Microsistemas, sales represented less than 25 per cent of what had been forecast in 1986. It should be pointed out that, like Microsistemas, firms in this medium-sized group (not a large economic group) had extensive technological experience in electronics (basically industrial electronic systems and equipment). Autorede concentrated on several types of point-of-sale, automatic teller machines and scanners intended mainly for the largest supermarket chain in the country.

The third project was that of ICSA, which by the end of 1988 was employing 60 people, 59 per cent of whom were engineers or technicians. A drastic reduction in state demand meant that exports became much more important than originally forecast (representing 48 per cent of sales during the first year of the promotion). This was the only case in which sales exceeded the original projection. In the second year of the project the company's employment and sales figures were higher than those projected for the sixth year. The firm belongs to Pescarmona, which has extensive experience in engineering industries and, in this regard, ICSA, like Microsistemas and Autorede, has experience in highly skill-intensive sectors. The main customer for ICSA's products (programmable logic controllers and microprocessor-based control equipment) is the largest company in the group (IMPSA) involved in producing large, metal-mechanical equipment. More and more electronic measurement and control equipment is incorporated into their products.

Despite problems these three projects revealed certain features that confirm marked differences between the industry that this policy was trying to promote and the rest of the local EC. For example, the total proportion of electronics engineers, systems managers, other professionals and technicians on the staffs of these three companies is several times higher than the proportion given for the whole of the local EC (nine times higher for Microsistemas, five for Autorede and eight for ICSA). The percentage of sales assigned to R&D and the exports/sales ratio are also much higher, as are investments in equipment.

4.2 The tariff policy of the EIP and its impact on the electronic capital goods industries[17]

One of the central features of the tariff policy was to redress (if only partially) the lopsided liberalization process which began in 1976. Throughout the EIP, objectives were much more ambitious than could be achieved. Although the tariff policy could not reduce the effective protection rate for scale-intensive and assembling electronics activities, it at least raised the effective rate in the skill-intensive sectors, especially in industrial electronics, dedicated information technology, medical electronics and short-run communication equipment manufacture. This was particularly important because, while the previous scheme only protected activities with comparative static and dynamic *disadvantages* (such as final assembly of consumer goods and high-capacity switching equipment), the EIP tariff system increased the effective rate of protection for activities with comparative static (intra-industry) and dynamic *advantages*, making effective rates more symmetrical (Nochteff 1994). These advantages began to reveal themselves during the four short years in which the tariff system was in force.

Between 1985 and 1988, electronic capital goods' share of total EC production increased very significantly: telecommunications rose from 8% to 31% (basically owing to the use of surplus capacity), computers rose from 6% to 12%, medical electronics from 0.3% to 0.6% and industrial electronics from 1.7% to 5%. In that period productivity rose by a minimum of 50%, in computers, to a maximum of 94%, in industrial electronics. As an aside, this performance shows that macroeconomic instability – although it had some effect – was not a major constraining factor for the Argentine EC, at least until 1989.

There was a sharp increase in the number of engineers, systems analysts, other professionals and technicians employed, and a reduction in the number of skilled and unskilled operators. In other words, the profile of skilled workers improved considerably, a change consistent with Argentina's intra-industrial comparative advantages (relative abundance of highly skilled labour and relative shortage of unskilled labour and capital).

This change in profile emanated from the recruitment attitude of the capital goods sectors, which in 1988 employed 85% of the electronics engineers, 98.3% of systems analysts, 65% of other professionals and 75% of technicians and, at the other extreme, 51% of unskilled operators.

Indicators suggest satisfactory technological development in these sectors. The proportion of R&D employees compared to the total number of workers was 3% in the whole of the EC, and 4.2% in the capital goods sectors (5.3% in industrial electronics). The same pattern could be seen in wages paid for R&D activities as a percentage of the total wage bill and production value. In line with this, the number of new patents based on in-house developments quickly grew. Some 85 per cent of these patents were generated by electronic capital goods companies. Technology transfer (TT) contracts for the period show that an increase in R&D activity in the capital goods sectors was accompanied by a search for outside technology. In this respect it is important to point out that TT contracts in the consumer goods sector were more closely tied to input suppliers than those in the capital goods sectors. Technology transfer contracts with input suppliers were 77 per cent of the total in consumer goods and 8.7 per cent in industrial electronics. The proportion of such contracts was also high (67 per cent) in the computer sector, mainly involving companies covered by promotion schemes which co-existed with the EIP scheme and which did not have any technological or production performance requirements.

Finally, export performance was also positive: between 1984 and 1988 telecommunication equipment exports increased from US$119 thousand to US$3.0 million, scientific and medical instruments from US$1.5 million to US$3.0 million and industrial electronics from US$545 thousand to US$4.9 million. Although these figures are modest, it must be borne in mind that the EIP never included an export-promotion system, that domestic sales were growing quickly at that time, and that export dynamism in these industries cannot be easily achieved in the short run, especially after ten years of decrease in exports. Computer exports were totally atypical. The increase from US$77.8 million to US$88.9 million was almost totally dependent on the achievements of the IBM Argentina factory, from which no conclusion can be drawn in relation to the effect of the EIP. All these production, technological and commercial performance indicators are consistent with an increase in the effective protection rate and the hypothesis that some of the electronic capital goods sectors had intraindustry comparative advantages, at least with respect to other EC sectors. On the whole they can be taken as rough indicators of increased external competitiveness.

To sum up, the analysis of the EIP's impact on the EC leads to the following conclusions:

1 Within a context of overall failure, the policy had positive technological and industrial effects during the period when it was partially applied.
2 These positive effects were not a consequence of unforeseen results, but materialized to a limited extent when resources shifted towards skill-intensive sectors which were not dominated by greenhouses/enclaves, thus reversing historical trends.

5 SUMMARY AND CONCLUSIONS

The development of the electronics and informatics policy (EIP), formulated between 1984 and 1985, exemplifies the implications of the exclusion of technological policies and related industrial policies from big business and government agendas. The EIP set up innovation and technological skills as central objectives, attempting to reform government and big business priorities, behaviour and policy style. The officials who formulated and implemented the EIP started with the implicit assumption that a reform of this scope could be made by an industrial and technological sectoral policy. The EIP experience proved the assumption wrong.

Analysis of the EIP reveals great differences between the policy formulated in 1984 and 1985 and that implemented between 1986 and 1990, suggesting the idea of two policies: the one that was formulated and the real one. The factors which explain the direct causes of these differences and which led to the breakdown of the policy were:

1 the persistence of the traditional, adaptive, exogenously driven behaviour of the Argentine industrial economy and of the policy priorities related to this behaviour;
2 the regressive restructuring which was begun in the mid-1970s; and
3 the behaviour of big business, especially its location in non-transitory monopolies which were not related to innovation, and its shift toward non-tradables and 'greenhouses'.

Both state structures and the style of government policies emerged from, but at the same time propagated, the classical 'economic restructuring' model. Never a priority, the technological and industrialization questions relating to the capacity for innovation disappeared from the state agenda. State reorganization in the mid-1970s accompanied the restructuring. For this purpose the state either transfered massive resources to economic 'greenhouses' or created sophisticated mechanisms to subsidize the operation of an oversized financial sector. However it was unable to draw up and administer scientific, technological or science-based industrial development policies, or policies focused on specialized suppliers, and was even less able to administer policies efficiently.

In line with the above, it should be noted that an implicit overall industrial policy continued throughout the period, from 1984 to 1988. Since at least 1976 this policy had been characterized by the actual reversal of the infant industry argument. Therefore, neither the political system, including the government, nor big business gave any priority to the EIP. On the contrary, after the first stage of formulating strategic outlines, both the scope and resources assigned to this policy were extremely limited. Because of this, the ability to solve the questions relating to its application was considerably reduced.

Even the large economic groups who participated in the promotion scheme prioritized other activities, as financial operators, commodity producers, civil engineering contractors, equipment importers, service providers and as lobbyists, instead of taking on riskier investments. Reformist officials were never supported by big business and political actors. This is consistent with the fact that the objectives, requirements and the very style of the EIP were quite different from those of other government policies (especially those of the immediately preceding years) and from the behaviour of big business. This difference is evident in virtually all of the EIP aspects such as: transparency and competition vs. lobbying and ad hoc concessions; performance-related subsidies vs. subsidies not related to performance; the granting of transitory, monopolistic advantages which were dependent on innovation vs. non-transitory monopolistic conditions not dependent on innovation; involvement vs. exclusion of small and medium-sized companies; and the protection of capital goods production vs. import of turnkey plants.

Looked at from this point of view, the experience of the Argentine EIP teaches us a lesson. In countries like Argentina, where the major economic agents, decision-makers and entire society are concerned with neither growth by industrialization, nor increased structural competitiveness, nor integration of technical progress into production, then conventional methods used everywhere else to develop industrial policy simply do not apply. In other words, in Argentina it does not seem possible to resolve problems of technology and industrial development without taking it into a larger context. The failure of the 'reformist' officials who initiated the EIP to attract the large economic groups so as to obtain the political resources they lacked is directly, although not solely, connected with the sectoral limits of the EIP. The benefits it offered, in combination with a demand for performance, could not be attractive if limited to a small, although dynamic, sector of total economic activity. This was the more so when the benefits offered in the rest of the economy were more attractive and performance demands were much lower. In retrospect, it is obvious that the large diversified economic groups which had – or could obtain – better offers of support with less stringent demands attached would not pay any attention to this sectoral option. It can be said that a policy which presented a limited and restricting choice to large economic groups who had before them an extensive menu of easy alternatives was almost certainly condemned to failure.

As regards the TNCs, it is necessary to distinguish between two situations. First, that of IBM, which was running an industrial operation in the country and for which the EIP was neutral or negative. EIP could not offer benefits to IBM industrial plant but could raise the tariffs on equipment imported for resale in the local market. As well, the EIP could effectively increase competition in the state market, which was led by IBM. IBM was opposed to the policy but its influence, although important, was not decisive. It is likely that if the policy had gone ahead, the confrontation would have become more acute, but this did not in fact occur. The EIP failed due to lack of support from the large Argentine economic groups and from the political system and also because of a general resistance from the environment, and not because of the deliberate opposition of a single identifiable actor. Some of the other TNCs, which did not have plants in the country, supported the EIP. They reasoned that if state purchases were redirected towards the promoted companies, then they could win a share of the market controlled by IBM. Nevertheless, these TNCs lost interest in supporting the EIP when evidence showed that (a) they would not gain a share in the market, (b) the 'reformist' officials were extremely weak and (c) their local partners (the large economic groups) would not commit themselves to the EIP. In this respect the Argentine experience was quite different from that of Mexico or Brazil. In the first place, the EIP, unlike the Brazilian Informatics Policy and the first version of the Mexican Policy, did not exclude TNCs, and in the second place it did not immediately push IBM on the one hand and the large Argentine economic groups, the other TNCs and the government on the other hand into direct confrontation.[18]

A second lesson to be learned from the EIP experience is that basing the assessment of a policy only on its formulation, without considering its effects, is misleading. This obvious lesson is worth pointing out since it has been ignored in the majority of the discussions of Argentina's industrial policy, especially with regard to electronics. The magnitude of any incentive must be measured, *ceteris paribus*, in relation to the significance of other incentives that are offered in one and the same economic system. Drastic judgements should not be made as to the insufficient effects of the promotion of the Argentine electronics industry, because it remains an open question whether the industry was actually promoted. A clear example of this is seen in the protection policy: there is not much sense criticizing the disadvantages of giving the capital goods electronics industry, and informatics in particular, a higher effective protection than the other industries because it is very debatable whether that ever occurred. The actual rate of protection for computer products does not seem to have been among the highest in Argentina in the period 1986–90. On the contrary, for some of the promoted production stages, the rate of protection seems to have been lower than the rate for cars, printing paper and television assembly. In general terms, it is essential to clarify which industrial policies were actually applied in the country in the mid-1970s. Everything points to the fact that it would be very difficult to maintain that, from 1976 onwards, competition developed through liberalization and deregulation. It is just as difficult to maintain that the electronics industry was really promoted in relative terms.

Despite the above, if we compare the performance of the capital goods electronics industries during the period 1984–8 and that of the assembly electronics industries, or even when comparing it with the whole of Argentine industry, we can confirm the hypothesis that regressive restructuring policies, which were explicitly geared to exploiting comparative advantages, acted against intra-industrial dynamic comparative advantages and even against static ones. Conversely, the EIP, especially the tariff system, seems to have worked, both implicitly and explicitly, in favour of those advantages.

Everything indicates that the factors that prevented the creation of the political, institutional and economic links necessary for the success of the informatics policy are the same ones that hindered the formation and consolidation of progressive restructuring for the Argentine economy. In this context, both the behaviour of the major economic agents and the priorities and weaknesses of the political system contrast with what happened in the international economy. Currently, comparative advantages, international competitiveness and the division of labour are to a large extent the result of joint strategies of the major economic and political actors. Governments are increasingly playing a decisive role in international competition, and this leads to a continuous confrontation between production systems, institutional models and social organizations in which private companies hold prominent places. However, companies comprise only one component of a network which links them to educational systems, technological infrastructure, management/employee relations, public and private institutional apparatus and financial systems. 'High technology neo-mercantilism' treats S&T and industries that bring technical progress as 'weapons for international competition'. This so-called technology 'neo-mercantilism' falls within the framework of intensified competition between countries and is linked with global competition. These countries organize national and regional systems in which competitive dynamics and industrialization are determined by the level of each community's capacity for technological, organizational, social and political change. In the light of all this, the behaviour of Argentina's economy, together with macro-economic policies (including those responsible for regressive restructuring), formed the most central constraints on the development of technological and industrial policies and, to that extent, to the dynamics and competitiveness of the economy.

Although this chapter deals only with the Argentine case, it is likely that the viewpoint taken, and the principal conclusions, are relevant to any discussion of the technological and industrial policies of other countries in the region, especially those geared towards the development of the EC. First, many of the characteristics of Argentina's EIP are common to the policies of the majority of countries in the region:

1 the late and narrow perception of the importance of electronics and informatics, not only in comparison to industrialized countries, but also to the most successful industrial latecomers;
2 the explicit objectives;
3 the contrast between the ambitious formulation and the poor implementation; and
4 the difficulty in achieving a minimum level of convergence between the policies which applied to the various EC industries, despite the fact that such convergence is generally considered essential.

Second, many of the environmental features are also common:

1 the rent-seeking behaviour of many major economic agents and their strong aversion to risk;
2 the industrialization model with minimal development of sectors promoting technical progress; and
3 the existence of protected industries, sustained by permanent income transfers, in which the bulk of capital formation is located.

These characteristics and behavioural patterns, reciprocally fed by the inability to open the 'black box' of technical progress, have brought stagnation or regression to production and real competitiveness, and leave many countries in the region only passive ways of breaking into the international economy.

Perhaps the most biting description of the attitude of Argentina's major social actors towards S&T has been given by one of Argentina's most important scientists:

> Nearly everybody sees national scientists as little more than a symbol of the state. As the country already had opera, museums, art galleries, zoos and international sports arenas, businessmen thought that things were not going so badly. Certain scientific work – and what can you say about a Nobel Prize? – gave a comforting touch of distinction which completed the picture. We researchers were then national status symbols. Whenever the President travelled abroad, Houssay and Fangio [Nobel Prizewinner and Formula One World Champion respectively] were part of his retinue. However, Argentina, which once boasted of being the 'granary of the world' was paying royalties to those countries to which its scientists were emigrating in order to feed its chickens a balanced diet. There were companies which were making plastic plates, toys and babies' potties which suddenly went bankrupt because Europe or the United States had invented a new monomer which they did not know how to polymerize. But if one were to suggest to them that they offer financial support to the university polymer laboratory to train people to master these techniques, they would have been scandalized and ended up importing machinery and contracting technical advice from overseas on terms and conditions laid down by the exporter.
>
> (Cereijido 1990)

NOTES

1 The author is especially grateful for the comments of Eduardo Basualdo, Roberto Bisang and Marcelo Diamand.

2 Ironically, it was an extreme right-wing military dictatorship with a neo-conservative, monetarist economic administration which, if we look at the results carefully, exercised a profound technological 'decoupling'.

3 The fact that investment fell below the levels necessary to cover capital depreciation, and that practically all investment was subsidized (Azpiazu and Basualdo 1987) confirmed this shift, especially if we bear in mind that the same industrial companies which did not invest in fixed assets were major holders of monetary assets (Damill and Fanelli 1988).

4 The difference between Argentine interest and devaluation rates was much higher than the difference between interest and inflation rates in industrialized countries. Therefore if one bought pesos with dollars, deposited them in the financial system and a while later bought back dollars, the profit, measured in dollars, was very high. It should be pointed out that for this to be possible it is not necessary for the real interest rate to be positive, because the devaluation rate can be much lower than inflation and interest rates, as long as trade and service imbalances are compensated by capital inflow.

5 The less diversified industrial firms could not offset drops in production and in the relative prices of some product lines with increases in others. Small and medium-sized companies did not have access to foreign credit or state guarantees.

6 It must be remembered that this was a military dictatorship and therefore, as usually happens in these political regimes, the lobbying power of the major companies increased while the feedback between the state and the other social actors decreased. This characteristic was very noticeable during the Argentine military dictatorship of that period, when the activities of practically all intermediate political and social organizations (including some of the employers' organizations) were suspended.

7 This is because, in a highly segmented credit market with virtually a permanent shortage of funds, industry was self-financing through prices, thanks to high tariffs. When protection fell, companies (mainly small and medium-sized) were left without finances, while the large economic groups and some TNCs, involved in sectors where the effective protection rate fell much less, but which also had access to foreign credit, improved their financial conditions. In addition, they paid a lower interest rate for loans in the local market. The real interest rate for each borrower is a positive function of the nominal interest rate and a negative function of the increase in the price of what he is selling. Consequently, the real interest rate can be negative for non-tradables (which can increase their prices during a period of exchange appreciation and commercial liberalization), positive for tradables (which cannot do this), and negative for those tradables that are protected despite the liberalization: i.e. for the large economic groups and some TNCs.

8 The same ones which benefited from indebtedness, from the sectoral bias of the liberalization and from the nationalization of the foreign debt (Azpiazu and Basualdo 1989).

9 This does not mean that the first group of sectors is not dependent on the stage of development or the consolidation of an industrial system and on a national system of innovation, but in relative terms it is much less dependent on those factors than the second group, as may be seen from its global geographical spread.

10 Many of these sectors, such as iron and steel and petro-chemicals, are in theory tradables but in Argentina they were (at least until 1992) minimally tradables, as they are exportable but difficult to import for various reasons. The most important reason, but not the only one, is the imperfection of the markets (Nochteff 1991b). For example, when the same large economic group partially controls the railways, toll roads, the production of steel and rolled laminates, gas and electricity, and the wholesale distribution and service to the public of at least one of these energy sources, it can set a very low price, actually a transfer price, on the energy it consumes and at the same time raise the prices paid by consumers or small and medium-sized firms. It can do the same with transport, thus basing its competitiveness on a transfer of income administered by the same large economic group. If in turn, as in the case in point, the monopoly sells to a large number of medium-sized companies, it is doubtful whether these could import steel, even if the monopoly's prices are not competitive, due to the minimum quantities required for shipments. Briefly, in this situation steel, which is in principal tradable, ceases to be so and the company is totally or

partially protected from overseas competition even if the nominal tariff is reduced (in such a case the real effective rate of protection may be very high even with a low nominal rate).

11 The description and analysis of electronics and informatics during the period 1984–8 and the behaviour of the electronics and informatics industry set out in this section are based on a research programme carried out between 1984 and 1992. The results of this research have been published in Azpiazu, Basualdo and Nochteff (1988, 1989, 1990a, 1990b, 1991a, 1991b) and Nochteff (1985, 1990, 1993). Except where a different source is indicated, the statements made in this paper are based on these publications.

12 In 1988, the composition of the Argentine electronics complex, measured by each segment's share of total production, was as follows: electronic consumer goods industry 35.6%, telecommunication equipment 31.3%, data-processing equipment 12.4%, cables 12.7%, electronic industrial equipment 5.0% and electronic medical equipment 0.6% (Nochteff 1993).

13 The CNI recommended a very wide spectrum of policies, on topics such as information technology and education, trans-border data flows (TBDF) and the software industry. In this article we will deal only with the policies relating to hardware production, as the others were not implemented.

14 During the import-substitution and liberalization periods, protection of consumer goods was much higher than capital goods; during the second stage some finished goods were protected through non-tariff barriers while very low or zero tariffs were set for sub-assemblies.

15 Regardless of absolute costs and endowments, Argentina's intra-industrial comparative advantage is highly skilled labour and its disadvantages are the cost of unskilled labour and the low capital endowment, shown in the low investment/product ratio (Nochteff 1993). The work mentioned shows statistically that the existence of greater symmetry in the effective tariffs inside the electronic complex between 1985 and 1988 revealed these advantages even faster, when measured by proxies for the competitiveness of skill-intensive activities.

16 Interviews with state officials and businessmen were conducted, mainly for the research projects mentioned as sources for this section, but also in the preparation of this chapter.

17 This section is based on research on EC performance, analysed in Nochteff (1993).

18 At the beginning of the policy there was also an 'expression of concern' by the US government, but this pressure was not repeated since on the one hand there were North American companies interested in the EIP (such as Unisys and Stride), and on the other hand the confrontation with IBM never turned into open warfare, although the company exerted successful pressure on several occasions to obtain tariff reductions on imported computers.

REFERENCES

Azpiazu D. and Basualdo, E. (1989) *Cara y Contracara de los Grupos Económicos*, Buenos Aires, Cántaro.

Azpiazu, D., Basualdo, E. M. and Nochteff, H. (1988) *Revolución Tecnológica y Políticas Hegemónicas. El Complejo Electrónico en Argentina*, Buenos Aires, Legasa.

Azpiazu, D., Basualdo, E. and Nochteff, H. (1989) 'La política industrial informática en Argentina', FLACSO, *Serie Documentos e Informes de Investigación*, no. 90, November, Buenos Aires.
Azpiazu, D., Basualdo, E. and Nochteff, H. (1990a) 'Industrial Policies and Industrial Restructuring: Lessons from the Argentine Informatics Sector', paper presented in the workshop 'Hi Tech for Industrial Development', IDS, Sussex University, June 20–2.
Azpiazu, D., Basualdo, E. and Nochteff, H. (1990b) 'Los límites de las políticas industriales en un período de reestructuración regresiva: el caso de la informática en la Argentina', in *Desarrollo Económico*, vol. 30, no. 118, pp. 151–72.
Azpiazu, D., Basualdo, E. and Nochteff, H. (1991a) 'Contexte Economique, Comportement des Acteurs et Politique Industrielle: Le Cas de l'Informatique en Argentine', in *Cahiers d'Economie Mondiale*, January, vol. 5, no.1, pp. 27–43.
Azpiazu, D., Basualdo, E. and Nochteff, H. (1991b) 'Dos décadas de comercio exterior de bienes electrónicos de la Argentina. Relevamiento y análisis', FLACSO, *Serie Documentos e Informes de Investigación* No 107, Buenos Aires.
Azpiazu, D. and Kosacoff, B. (1987) 'Exportación de manufacturas y desarrollo industrial. Dos estudios sobre el caso Argentino', CEPAL, *Documento de Trabajo* no. 22.
Bastos, M. I. (1992) 'The politics of Science and Technology Policy in Latin America. Draft of Analytical Framework', Maastricht, UNU/INTECH.
Basualdo, E. (1987) *Deuda Externa y Poder Económico en la Argentina*, Buenos Aires, Nueva América.
Basualdo, E. and Fuchs, M. (1989) 'Nuevas Formas de Inversion de las Empresas Extranjeras en la Industria Argentina', Buenos Aires, CEPAL Working Paper no. 33.
Calcagno, E. (1982) *Los Bancos Transnacionales y el Endeudamiento Externo en la Argentina*, Santiago de Chile, Cuadernos de la CEPAL.
Canitrot, A. (1982) 'Teoría y Práctica del Liberalismo. Política Antiinflacionaria y Apertura Económica en Argentina 1976–1981', in *Desarrollo Económico*, vol. 21, no. 82, pp. 131–89.
CEPAL (1992) 'El comercio internacional de manufacturas de la Argentina 1974–1990. Políticas comerciales, cambios estructurales y nuevas formas de inserción internacional', Working Paper, Buenos Aires.
CEPAL (1987); 'Perfiles Industriales de América Latina', in *Industrialización y Desarrollo Tecnológico*, Informe no. 4, División Conjunta CEPAL/ONUDI de Industria y Tecnología, Santiago de Chile.
Cereijido, M. (1990) *La Nuca de Houssay*, Buenos Aires, Fondo de Cultura Economica.
Damill, M. and Fanelli, J. M. (1988) *Decisiones de cartera y transferencias de riqueza en un período de inestabilidad económica*, Buenos Aires, CEDES.
Dorfman, A. (1970) *Historia de la Industria Argentina*, Buenos Aires, Del Solar.
Fajnzylber, F. (1983) *La Industrialización Trunca de América Latina*, Mexico D.F., Nueva Imagen.
Flamm, K. (1988) 'Trends in the computer industry and their implications for developing countries', Stanford University and The Institute of the Americas.
FMI (1986) *Argentina, incentivos fiscales para el fomento del desarrollo*, Washington.
Frenkel, R., Fanelli, J. M. and Rozenwurcel, G. (1990) 'Growth and structural reform in Latin America. Where we stand', Buenos Aires, CEDES.

Herrera, A. (1989) *La Revolución Tecnológica y la Telefonía Argentina*, Buenos Aires, Legasa.

Katz, J. and Kosacoff, B. (1989) *El proceso de industrialización en la Argentina: Evolución, retroceso y prospectiva*, Buenos Aires, CEPAL/CEAL.

Nochteff, H. (1985) *Desindustrialización y retroceso tecnológico en la Argentina. La industria electrónica de consumo 1976–1982*, Buenos Aires, FLACSO/GEL.

Nochteff, H. (1990) 'Argentina: crisis económica, reestructuración industrial y comportamiento de actores', in *Cono Sur*, vol. IX, no. 3, pp. 11–27.

Nochteff, H. (1991a) 'Paradigma tecnológico, actores sociales y control de la interdependencia', in *Espacios de Crítica y Producción*, Facultad de Filosofía y Letras, University de Buenos Aires, no. 10, pp. 36–43.

Nochteff, H. (1991b) 'Reestructuración Industrial en la Argentina: regresión estructural e insuficiencia de los enfoques predominantes', in *Desarrollo Económico*, vol. 31, no. 123, pp. 339–58.

Nochteff, H. (1993) 'Comportamiento económico y politicas tecnológicas en la Argentina', *Revista Ciclos en la Historia, la Economia y la Sociedad*, Instituto de Investigaciones de la Historia Económica y Social, Facultad de Ciencias Económicas, Universidad de Buenos Aires.

Nochteff, H. (1994) 'El complejo electrónico en la Argentina. Evolución reciente y competitividad', Secretaría de Programación Económica del Ministerio de Economía, Buenos Aires.

Nochteff, H. *et al.* (1988) 'Reestructuración Industrial y Competitividad Internacional: Propuestas para el Ecuador', Executive Report, Report no. 1, Project SI/ECU/88/060.

Sábato, J. F. (1988) *La Clase Dominante en la Argentina Moderna. Formación y Características*, Buenos Aires, CISEA/GEL.

Shapiro, H. and Taylor, L. (1990) 'The state and industrial strategy', in *World Development*, vol.18, no. 6, pp. 861–78.

Wade, R. (1990) *Governing the Market. Economic Theory and the Role of Government in East Asian Industrialization*, Princeton, NJ, Princeton University Press.

World Bank (1987) *Argentina. Industrial Sector Study*, Washington, DC, The World Bank.

World Bank (1989) *The Impact of the Industrial Policies of Developed Countries on Developing Countries*, Washington, DC, World Bank and International Monetary Fund, Development Committee.

6 The political economy of technology development: the case of the Brazilian Informatics Policy

Fabio Stefano Erber[1]

This chapter is divided into three sections. In the first the history of the Brazilian Informatics Policy (BIP) is recounted, from its origins to its end, focusing on the evolution of its objectives, instruments and the political alliances which supported and opposed the policy. It is based on the literature (notedly, Bastos 1992; Dantas 1988; Gaio 1992, Piragibe 1985; Schmitz and Hewitt 1992), on interviews with policy-makers and on first-hand experience. The second section is more analytical and tries to explain the fate of BIP. It begins by restating the 'model' of BIP and then proceeds to discuss its assumptions, first the technical and economic assumptions and the problems it met in its attempt to manage the technological gap and then the values and interests at stake and how it failed to retain social support. The last section comments on some of the results of BIP.

1 THE BRAZILIAN INFORMATICS POLICY: A CAPSULE CHRONOLOGY

Historical periodization of a policy necessarily involves a degree of arbitrariness because of the cumulative nature of the processes underlying the policy. As an example, it may be claimed that BIP originated in the early 1950s, when the ITA (Instituto Tecnológico da Aeronáutica – Air Force Technology Institute) started its electronics engineering course. This provided a considerable part of the technical expertise upon which the policy was grounded and where a significant number of cadres responsible for the policy was formed. Because we are dealing with a policy under an evolutionary approach, the main criteria adopted here are the life-cycle and institutional aspects.

Following these criteria, we may say that the BIP's life-span covers twenty years: from the creation of the Special Working Group (GTE – Grupo de Trabalho Especial) jointly established by the Navy and the

National Economic Development Bank (BNDE) in 1971 to the new Informatics Policy Law of 1991, which drew BIP to a close. Such history may be subdivided into three periods, detailed below.

1.1 Infancy: from the GTE to CAPRE, 1971–5

During the 1950s and 1960s the Brazilian state (mainly the federal government but also the state of São Paulo) fostered the development of technical and scientific capabilities in fields related to BIP by supporting local academic institutions and by providing training abroad. However, this support lacked any specific industrial purpose – the use of the resources thus developed was left to the market. The demand for electronic equipment was supplied either by imports or by local production by subsidiaries of multinational companies. Consequently, employment of skilled personnel was directed mainly to selling and using equipment, with a minority being absorbed by the academia. State agencies were major users of such equipment.

One of the main sources of funding for graduate education and research was the BNDE, the main industrialization financial agency of the country. Although the bank's department for science and technology (FUNTEC) was marginal to the agency, institutionally and resource wise, it became an important breeding ground for science and technology policy-makers. In 1968 the department had floated the idea of developing a computer prototype in the local universities. Independently, the Navy had adopted in 1969 a policy of fostering the local production of electronics equipment used for its vessels and airplanes. The purchase of British frigates equipped with computers brought the two agencies together, under a Special Working Group (GTE 111) which led to a project aimed at designing, developing and producing computers for naval use. As a consequence, in 1974 a company (COBRA – Computadores Brasileiros SA – Brazilian Computers) was established to produce such equipment – a joint venture between the Navy supplier of computers for its frigates (Ferranti), a local supplier of other electronic equipment to the Navy (Equipamentos Eletrônicos) and the Bank. A second company was planned, under the same tripartite model, to produce commercial equipment, but it never took off.

In the same period the tripartite model, combining Brazilian state and private capital and foreign enterprises, the latter acting mainly as suppliers of technology, was successfully used for the establishment of the petrochemical industry in the northeast of the country. The model was conceived in order to ensure a coexisting private and national majority control of the capital of the new enterprises, while providing the enterprises access to foreign technology.

Simultaneously, other electronic data processing equipment was developed by state agencies, mainly for processing fiscal data and for cryptography, and by some universities, mainly for academic purposes. In 1971, the Planning Ministry revamped a small agency dedicated to financing pre-investment studies (FINEP – Financiadora de Estudos e Projetos – Financial Agency for Studies and Projects) as a development bank for science and technology, initially run by bureaucrats from BNDE's FUNTEC. The support of graduate education and research in computer sciences and the supply of computers to universities soon became a priority of FINEP.

At the user end, the Ministry for Planning and Co-ordination established in 1972 a Commission for Electronics Processing Activities (CAPRE – Comissão de Atividades de Processamento Eletrônico) in order to rationalize the use and purchase of computers by the federal government as well as the training of personnel. In 1975, following the severe foreign currency restrictions due to the oil shock, CAPRE was empowered to control imports of computers and parts and components, which had become main import items, increasing 600 per cent between 1969 and 1974 (Piragibe 1985).

CAPRE was staffed by technical cadres originating from the bureaucracy and from the academia. In turn, CAPRE officers helped to organize scientific societies and meetings related to computerization and the associations of electronics data-processing professionals (e.g. systems analysts, programmers, etc). The three groups – bureaucrats (civilian and military), academics and professionals – formed BIP's hard-core, providing its concepts and political support. Given the authoritarian nature of the regime, only the former were audible within the state but the latter two were quite vocal publicly.

Although the First National Development Plan (1972/74) and the First Plan for Science and Technology (1973/74) listed the computer industry among their priorities, singling out minicomputers, and in spite of the initiatives mentioned above, there was not a policy for the sector nor an agency in charge of it. None the less, the main political and institutional building blocks of BIP had been laid down.

1.2 From youth to maturity under bureaucratic rule: from CAPRE to SEI, 1976–83

The Brazilian economic history of the second half of the 1970s is dominated by the federal government Second Development Plan – a major industrialization thrust aiming at, on the one hand, completing the Brazilian industrial structure by large investments in the industries producing intermediary products and capital goods and, on the other hand, changing the energy matrix, by replacing gasoline by alcohol and by complementing the large investments in hydro power with an ambitious nuclear power programme, jointly developed with the Federal Republic of Germany. Underlying the plan was the idea that Brazil was an 'emerging international power'. Although the plan was led by the state, its aim was to strengthen the nationally owned enterprises while avoiding as much as possible antagonizing the multinational companies, especially those already established in the country.

The development of local technological capability, including innovation skills, was an important part of the plan's design, since imports were associated with undeveloped capability. Accordingly, FINEP's resources for developing the research and graduate education system and for granting subsidized loans to technology projects of national enterprises were expanded. State enterprises were instructed to favour locally designed products in their purchases and they greatly expanded their own R&D centres. At the same time, stricter rules for importing technology were adopted, often conditioning imports to local technological efforts.

Parallel to the main thrust of the plan, which aimed at completing the structure of production of the second industrial revolution, the government policy-makers established a set of policies aimed at high-tech sectors, the vectors of the Third Industrial Revolution: telecommunication, aeronautics, armaments, nuclear power and, last but not least, informatics. Such sectoral policies had two elements in common: the emphasis on local technological capabilities (including innovation) and the control of the sector by national enterprises.

Within this context, in the beginning of 1976 CAPRE was restructured, gaining a council composed of high officers of several ministries and empowered to define a national informatics policy. The policy, stated in the middle of the year, had five main objectives:

1 to achieve technological capability to design, develop and produce electronic equipment and software in the country;
2 to ensure that national firms hold a prominent position in the national market;
3 to create opportunities for the development of the informatics parts and components industry;
4 to create jobs and, especially, more qualified employment for national engineers and technicians; and
5 to generate a favourable balance of payments for informatics products and services.

Retaining its power to control imports, which were necessary to manufacture computers in Brazil, CAPRE was also empowered to analyse local manufacturing projects. However, it lacked any positive inducement policy instrument, such as credits or fiscal incentives.

Within the sector, priority lay with the segment of small systems, mini and microcomputers and their peripherals, which had not yet been produced in the country and to which the efforts to develop national technology had to be directed. For larger systems, locally produced by subsidiaries of multinational firms, especially IBM and Burroughs, the emphasis was laid on rationalizing the investments by using the resources available.

Under the leadership of BNDE, the state took control of COBRA's capital and the company was turned around to become a supplier of commercial equipment, mainly for the banking system. A consortium of private banks took an equity interest, followed by the two other main federal state banks (Banco do Brasil and Caixa Economica Federal). Although it used licensed technology to produce its banking automation equipment, COBRA continued to invest heavily in its own projects, developing the small systems concepts originated from the GTE. However, an important by-product of this decision was the reduction of the Navy's support of the policy.

Simultaneously, IBM announced its intention to produce its minicomputer/32 locally. It was widely recognized that such production would kill the policy. As a consequence, battle lines were drawn: on the foreground, on one side IBM and other multinational subsidiaries and on the other academia and professional communities. In the background, the government was deeply divided and subject to strong internal and external pressures.

Such conflicts, which lasted throughout 1976 and 1977, led to a bid to manufacture minicomputers in Brazil. Projects were assessed by CAPRE according to five criteria:

1 local content of manufacture in terms of components and employment;
2 technological 'openness', giving higher priority to projects that involved greater disclosure of imported technology and greater local technological development;
3 control of the internal market in order to avoid excessive concentration in the hands of a single firm;
4 local equity control; and, finally,
5 balance of payments conditions.

Implicit, there was a belief that the import of technology would be a once-and-for-all affair and that next generations of the same type of equipment could rely on locally developed technology.

Fifteen projects were submitted to CAPRE, out of which three were selected at the end of 1977. All three were based on imported technology with the appropriate 'openness' clauses. None of the technology suppliers was an industry leader. In fact, the latter, through their subsidiaries in Brazil (IBM, Burroughs, HP, NCR, Olivetti and TRW) chose to participate in the bidding without local partners and with scant attention to the other conditions of the bid. Out of the pressure exerted by segments of the intelligentsia and the bureaucracy and the rigidity of the multinational leading firms, the market reserve was born.

The market reserve for locally controlled firms was applied to minicomputers and smaller systems only. Combined with import restrictions it implied that the segment of larger systems was *de facto* reserved to the subsidiaries of multinational firms already established in Brazil – mainly IBM and Burroughs. Because of its product mix the other main foreign supplier, Olivetti, stood to lose more from the policy and eventually withdrew from the sector. Although they were barred from the fastest-growing market segment, the other two, and more clearly IBM, henceforth changed strategy, refraining from explicitly attacking the policy and attempting to circumvent it by playing upon the definition of the limits of the policy, producing computers of 'medium' size. However, other US companies, such as Data General, totally excluded from the Brazilian market, started pressing the US government to take action against the policy.

In 1979, with the change in government, a new actor came to the fore in the informatics policy: the national security community. Since the mid-1970s the National Security Council and the Foreign Affairs Ministry had been involved with informatics projects related to cryptography, developing equipment and software locally, which led to the creation of a specialized firm. Such experience strengthened their perception of the weakness of the Brazilian electronics industry. Considering electronics strategic for the objectives of national security, interpreted at the time in wide embracing terms, they considered CAPRE's policies weak and limited in terms of the range of electronic activities, and they viewed with suspicion its involvement with academia, a traditional focus of resistance to the military regime.

Holding very strong powers within the government, the Security Council, supported by the Foreign Affairs Ministry, sought the control of the informatics policy. This bid was largely unopposed by the economic ministries, which were at the time concerned mainly with the foreign adjustment of the economy and with controlling inflation. As a result, CAPRE, where the majority of the council was held by civilian ministries, was replaced by a Special Informatics Secretary (SEI – Secretaria Especial de Informática) attached to the National Security Council.

SEI immediately laid down its directives of the National Informatics Policy. Contrary to many expectations, the objectives of local technological and industrial development under national control were maintained. In fact, the directives comprise one of the few official documents in which the market reserve for national firms is explicitly acknowledged. Moreover, the range of activities covered by the policy was broadened so as to include software, national networks for data communication, and all components of informatics products and services.

SEI had three instruments to implement its directives: the control of imports, the granting of permission to local manufacture, and the supervision of purchases by state enterprises and agencies. However, in the latter case it was never able to develop a comprehensive policy for the multifarious Brazilian state. In order to foster local technological development, SEI established at the end of 1982 a research centre (Centro Tecnológico para Informática – Technological Centre for Informatics) with four areas of concentration: computerization, process automation, instrumentation and microelectronics.

Over the next three years, following its directives, SEI defined through its Normative Acts policies and participants in the areas of computers and peripherals, industrial automation equipment, electronics instruments, microelectronic components and software. However, progress in the latter two, the top priority areas, was slow, albeit for different reasons. In microelectronics there were two large industrial groups interested but they demanded fiscal and credit incentives which lay outside the pale of SEI and which the economic ministries were unwilling to give. As a result, investments were procrastinated while, internationally, the minimum scale of plants grew by leaps and bounds. For software, although SEI imposed import controls and conditioned the approval of manufacturing projects for general purpose microcomputers to the adoption of locally developed operating systems, such measures proved to be ineffective against copying and smuggling of imported software.

Moreover, SEI's attempt to establish a coherent policy for the electronics complex was foiled at two strategic points – telecommunications and consumer electronics. The first sector was ruled by the Communications Ministry, which followed an independent policy, allowing products manufactured by firms which were nationalized subsidiaries of multinational companies. The technology was either supplied by the former parent companies or developed by the National Telecom Research Centre. The second sector was located in a Free Trade Zone in the Amazons region, under the auspices of the Interior Ministry, and its firms operated mainly as users of foreign designs and assemblers of imported components, directing their products to the internal market only.

Lacking any positive inducement policy mechanism, the efficacy of SEI's policies was further undermined by resistance from the economic ministries (Planning and Finance), where the objective of local technological development had become a remote priority, as witnessed by a decline in the funding of FINEP and the reduction of BNDES's support to the policy. The same applied to COBRA, the state enterprise which had led the market technologically in minicomputers, and which was left to languish in terms of mission and resources. The lack of financial incentives for investments especially affected the development of the strategic industry of microelectronics components, the core of the electronics complex, signalling to the firms involved a limited government commitment to the policy.

SEI's policies were obviously supported by the many industrial firms that had entered the market. However, the policy inevitably brought it into conflict with foreign subsidiaries previously occupying the Brazilian market, which often opted to leave the country altogether or to supply it via local licensors. Users of equipment also often resented the restrictions imposed on the supply and the inevitable higher costs of infant goods. Moreover, the military nature of the Secretary estranged the academic community and parts of the civil bureaucracy which had previously supported the policy. Even the main beneficiaries of the policy – the local companies which arose under the umbrella of the reserved market – often resented the restrictions placed on importing components and the power wielded by the bureaucracy. Hard-core supporters of the policy also complained about the 'laxity' of SEI's criteria about technology imports, putting a negative premium on the firms which invested more in the local development of technology.

With the end of the military regime in sight, the civilian supporters of the policy, especially the academics, the professionals and the industrialists deeply committed to local technological development, sought the backing of Congress to pass a law giving the policy appropriate political legitimacy. At the same time, concerned with international pressures and wishing to establish the policy on firmer ground, the military backers of the policy within the government also submitted a bill to Congress, thus, refraining from using the power of the Executive to pass a decree.

1.3 Maturity and decay under the law: 1984–1990

In 1984 the Brazilian Congress passed by unanimous vote the Informatics Law. Symbolically, the date was the same as that thirty years before when the law creating the oil monopoly had been voted – possibly the single major victory of the country's nationalist faction. In essence, the law was a ratification of CAPRE's and SEI's policies. It confirmed the prime objective of industrialization-cum-local technological development and the priority given to locally controlled firms. In this latter aspect it went a step further and defined 'local control' so as to involve technology – a feature later embedded in the definition of a 'national enterprise' in the Constitution of 1988. Similarly, the scope of the policy, stemming from the definition of 'informatics', covered the whole electronics complex.

Institutionally, however, the law led to a break in the previous mould, in which the policy was defined by the Executive alone. Now Congress had to approve the three-year Informatics Plans (PLANIN); furthermore the council in charge of presenting the plans to Congress and taking the main decisions regarding the policy implementation included a strong representation of the civilian institutions supporting the policy (CONIN – Conselho Nacional de Informática e Automaçäo – National Council for Informatics and Automation). SEI was retained as the executive secretariat of the Council.

The instruments of BIP were also nominally increased, although no mention is made of a market reserve. SEI retained the power to control imports, albeit for a period of eight years, and gained the management of some fiscal incentives, especially for microelectronics and software. The creation of a special fund for financing R&D expenditures was vetoed by the President but CTI was confirmed as a policy instrument under the jurisdiction of SEI. COBRA, however, was left under the control of the state banks, the private consortium having left some years before, as the participating banks developed individual informatization strategies and acquired interests in other informatics companies or established subsidiaries to supply banking equipment. Although the importance of state purchasing power as an industrial and technological policy instrument was recognized, no mechanism of co-ordination was established. In practice, since the fiscal incentives were very limited, the instruments of the policy remained the same as before the law.

A major effect of the law was to impart a much stronger legitimacy to the policy, strengthening its power. However, the unanimity of the last vote in Congress was deceptive. In fact, BIP's two main attributes – development of a local innovation capability under control of national enterprises – made it intrinsically conflictive.

The period of preparation of the bill and the debates in Congress showed that opposition to the policy was not exclusive to politicians traditionally aligned with foreign interests but came also from a wide range of actors who felt their beliefs and interests threatened. This arc encompassed objectors to the principle of state intervention through to multinational companies active in the electronics complex with, in between, officers from the economic ministries; local enterprises operating essentially with imported technology; and multinationals from other sectors, fearful of the extension of the informatics policy to other areas.

As a result, the final version of the law contained several compromises, the most important being the limitation of SEI's power to control imports for eight years. Nothing was said about what would happen the day after. Some interpreted the deadline as the end of the market reserve and others argued that the control could be taken over by another agency (CACEX), which controlled the rest of Brazilian imports. The press, which strongly opposed the policy, supported the first view, which tended to become dominant.

When the new civilian government came into power in 1985, the informatics policy was entrusted to the newly created Ministry of Science and Technology but telecom and consumer electronics remained under different ministries. Moreover, the politicians in charge of the three ministries held diametrically opposed views about the policy, as was reflected within CONIN. Policy co-ordination within the electronics complex was rendered all but impossible. Support from the economic ministries was, at best, lukewarm and the retrenchement of the military power implied a weakening of the policy.

Outside the government, there was a backlash of the 1984 decision: while the supporters of the law demobilized their forces, the opposers strengthened theirs with a mounting press campaign. Since BIP provided a model which could be used for other industrial sectors, the trenches gained new participants, especially from other high-tech sectors still undeveloped in the country – biotechnology and fine chemicals. On both sides the participants had the same origins as in informatics, but since the strength of the opposers (e.g. multinational firms from the pharmaceutical industry) was much greater than that of the supporters (e.g. small Brazilian firms producing fine chemicals), the net result of the proposals to expand the scope of the policy to other sectors only weakened it further.

Against this background of internal forces came a strong external pressure: in September 1985 the US government announced the start of an investigation into the Brazilian Informatics Policy and the possibility of economic retaliation if discriminatory or unfair trade practices were found against US interests. The possibility of restricting Brazilian exports to the US market, of sectors (e.g. orange juice producers) which had nothing to benefit from BIP increased the private sector opposition to the policy. Within the Brazilian government, the occurrence of a conflict at a time of balance of payments restrictions and intricate foreign debt renegotiations was anything but welcome at the economic and foreign affairs ministries.

In spite of this mounting pressure, the Brazilian executive and legislative branches refused to change the Informatics Law and, in order to stem criticisms, SEI was reorganized in 1986 so as to gain greater administrative efficiency. None the less, a process of deterioration of the policy got under way, especially from 1987 onwards, as the more nationalistic faction of the ruling party weakened and the economic conditions of the country worsened.

Thus, CONIN gave way to US pressures in several instances of import restrictions and Congress passed a Software Bill which was a clear compromise with US demands. This was a major turning point in the policy because of the growing importance of software for the sector, and the limited protection provided by the bill to locally developed software. The core of BIP was undermined. Moreover, the President of the republic stated to the press that informatics was a 'special case', killing the expansion of BIP to other sectors, and that he would not repeat his 1984 vote favouring the Informatics Law he had given as a Senator. Following this, the Presidency took several administrative measures, e.g. regarding the control by SEI of imports for the assembly of electronic products in the Amazon Free Trade Zone and the classification of enterprises as 'national', which gave clear signs that the Executive was withdrawing its support from the policy.

Less obviously, but not less effectively, the Ministry of Finance put a low ceiling on SEI's foreign exchange allowance. The imports of electronic products depended on such quota and were restricted accordingly, creating considerable strain between SEI and users of such products, with the former taking the blame. Moreover, a blind eye was turned to the smuggling of electronic products, which reduced the market for locally produced goods and services and jeopardized the results of projects aiming at local technological development.

Possibly anticipating the demise of the policy, some of the leading Brazilian informatics firms strengthened their technological links to foreign suppliers and laid the ground for future joint ventures. At the same time they used smuggled components massively, undermining the core of BIP and making its end a self-fulfilling prophecy. Such strategies were deeply resented by the enterprises more committed to the technological objective of the policy, causing a rift in the informatics entrepreneurial community.

Under such pressures BIP's policy-making apparatus lost all initiative. SEI, accused of rigidity, ended up by condoning entrepreneurial strategies, such as those mentioned above, which ran counter to the heart of the policy and CONIN was unable to present Congress with a strong Second Plan for the sector, leading to the extension of the First Plan. The absorption in 1989 of the Ministry of Science and Technology by the Ministry of Industry, then run by a notorious opponent of BIP, put an institutional lid on the policy.

Therefore, at the end of 1980s, the Brazilian Informatics Policy in its pristine form was agonizing. However, it was not dead yet and, besides the entrepreneurs of the sector, it could still rally support from important segments of society, especially the intelligentsia and the bureaucracy. Among such supporters there was a consensus, made explicit during the preparatory discussions of the Second PLANIN, that the policy should and could be revived, albeit with the introduction of major changes. Such modifications should lead to a greater product selectivity, restricting its application to a limited range of products. Local manufacture should increase import content so as to profit from external technological developments and should aim at cost reductions and at exports. It was hoped that if such changes were introduced, the policy could regain internal political legitimacy and retain its main pillars – the commitment to local technological development and the fostering of locally controlled firms. However, the presidential elections held at the end of 1989, when the economy was bordering on hyperinflation and policy-making capability was reduced to a minimum, led to the victory of a candidate explicitly committed to a reduction in state intervention and an opening of the economy to imports and foreign capital.

The supporters of BIP, especially the academics, tried to rally public support for the policy, attempting to revive the Brazilian Informatics Movement (MBI) which had been the mainstay of the 1984 campaign, but with very little success. ABICOMP, the national producers' association, preferred instead futile attempts to negotiate with the executive branch, ignoring that the past alliance with the bureaucracy was dead. Committed to terminating the policy, the executive branch adopted in 1990 administrative and legal procedures this end, such as withdrawing the Second PLANIN Bill from Congress, reducing the number of products subject to import controls, and extinguishing SEI. Appropriately, the last battle was fought in Congress and a new Informatics Law was passed in 1991, after a strong press campaign against the old policy, and especially against congressmen who supported it.

The new law changes the concept of 'national enterprise', reducing it to a requirement of a majority of capital, which allows for joint ventures in which technological control is in the hands of the foreign partner. CONIN was downgraded and its composition altered, reducing the weight of the groups previously supporting BIP in favour of its opponents. SEI was extinguished and replaced by a Department (DEPIN) of a new Ministry of Science and Technology, which is virtually powerless.

The law replaces import administrative controls by tariffs, which will decline with time. By 1994 tariffs will range from 35 per cent for finished goods to nil for components not produced in the country. As positive inducement mechanisms the law envisages fiscal incentives for local production and R&D activities, as well as the use of the state purchasing power. However, none of these mechanisms has yet been implemented.

Buffeted by the combination of general recession and policy changes, the industry went into a deep crisis, of which there is no end in sight: as compared to 1989, it is estimated by DEPIN that in 1992 the local industry had reduced net earnings, employment and R&D investment by, respectively, 47%, 60%, and 69%. So far the worst affected were producers of components and peripherals. Suppliers of banking automation equipment and of computers have fared relatively better – the former because they have a leading edge on technology combined with a captive and rich market and the latter by virtue of associations with foreign companies. Such partnerships have led to the abandonment of all plans of local technological development and entail no small risks for the survival of the local firms when import liberalization comes into full stride.

Although it is beyond the scope of the present essay to develop more fully conjectures about the future of the electronics industry in Brazil, it is clear that a chapter of the country's science and technology and industrial policy has come to an end.

2 OBJECTIVES, SCOPE, INSTRUMENTS AND SOCIAL SUPPORT

2.1 Objectives and scope

BIP was an ambitious project, to say the least. It aimed at creating a high-tech sector – first an industry and later the whole electronics complex – in a developing country, endowing such industry with the full range of technological capabilities, from research to marketing, all under the control of national entrepreneurs. The two attributes – technological self-reliance and national enterprises – made up its *diferentia specifica*, carrying it a step further than the traditional import-substitution policies, which relied almost exclusively upon imports of technology, especially for innovation activities, and upon subsidiaries of international companies.

The two attributes were organically linked: it was correctly assumed that the multinational subsidiaries would not invest in research and development in Brazil, having the option to do so nearer their parent companies, in locations where they enjoyed strategic and systemic advantages. Therefore, only national companies would be a vehicle for achieving a relative innovation capability. As shown by other sectoral experiences (e.g. petrochemicals), joint ventures with international companies where the latter controlled technology did not lead to the transfer of technological innovation capabilities to the local company. Therefore, wholly locally owned companies were necessary. If madness it was, there was method in it.

However, the two attributes were never unqualified. Technological self-reliance was never confused with autarky: technology imports were always an important part of the strategy, as the starting point for local production, although it was assumed that, with time, they could be replaced by local alternatives. As originally envisaged, it was assumed that having mastered the innovation capability for a family of products (e.g. minicomputers), based on a combination of technology imports and endogenous investments, the local firms would be able tò prescind from further technology imports once they could follow the international frontier.

This process would be 'horizontally' repeated for new product families (e.g. microcomputers), widening the scope of the policy, but it was understood that this scope was restricted by economic and technological factors (e.g. mainframes were excluded). None the less, the technological interdependencies which are characteristic of electronics industries, making them an 'industrial complex', led to a substantial widening of the scope of the policy.

In other words, the policy objective, in its most radical version, was eventually to close the technological gap for a selected but growing range of families of products. It was a process of import substitution in which local manufacturing and innovation capabilities were sequentially developed. Protection against imports – of products and of their built-in design – was regarded as essential to such purpose. As originally envisaged, such protection should be coupled with positive incentives as venture capital, credits, tax reductions and state procurement. Thus, other measures besides import controls were needed to reduce risks and costs and to accelerate the catching-up process. However, as already pointed out, the second part of the policy instruments was never fully set up, retarding the closing of the technological gap.

The other attribute was qualified too: national enterprises were supposed to take over only the product fields not previously occupied by multinational subsidiaries. This strategy, designed to come to grips with economic and technical realities of the computer sector as well as to reduce the policy conflicts, was applied mainly during CAPRE's rule. As the policy broadened its scope, the 'empty space' approach become more difficult to follow, eventually leading to the exit of some multinational companies as producers in Brazil.

2.2 Managing the technological gap: resources, needs and time

An assumption was critical to BIP's model: that the national enterprises would invest the necessary amounts to absorb the imported technology and develop their own technology. The amount of investment required was considerable, in view of the fact that the firms had no previous electronics experience. Such investment involved high risks too – technical, economic and financial – as well as a long time horizon. BIP's strategists clearly underestimated the speed and intensity of technological change in the electronics industry and, consequently, the threshold of resources necessary for catching up with the international technological frontier and remaining there, while, at the same time, absorbing the imported technology in other product ranges.

The threshold was further raised and the nature of necessary technical resources revolutionized by the emergence of the microcomputer and the changes in microelectronics. At the beginning of BIP, its policy-makers correctly identified the minicomputer as a window of opportunity and emphasized product design skills. However, the coming of the microprocessor and of the microcomputer with open architecture transformed the fastest-growing product of the industry into a commodity, placing the emphasis on production and marketing skills. Moreover, these skills changed as well. Production moved from batch assembly to automated mass production and the main users changed – from large and medium enterprises where purchases were made putting a premium on technical characteristics of the product, to small enterprises and households, where price and user-friendliness were the main considerations.

In other words, the international frontier had not only moved forward but had also changed its shape and the gap changed accordingly. However, BIP reacted slowly to this momentous transformation, retaining the emphasis on product design and not paying enough attention to production techniques and economics and even less to marketing. This was one of the important causes of BIP's downfall. We will return to some possible reasons for this slow reaction a little later.

The fact that electronics is a fan-shaped industrial complex, composed of several industries catering to different markets (e.g. data processing, telecom, entertainment) but made interdependent by a common technological basis, provided by microelectronic components and software, raised the above-mentioned threshold and widened the gap between resources and needs. SEI explicitly embedded the concept of the industrial complex into BIP but failed to gain control of the technology and industrial strategy of two industries which are critical for the electronics complex –

telecoms and durable consumer goods. As a consequence, economies of scope across industries and economies of scale for components and software were reduced. As previously discussed, the insufficiency of BIP's instruments (mainly import controls) contributed to stymie the development of software and components. As a result of this negative synergy within the electronics complex, the industries under BIP's aegis had their costs and risks augmented, especially when compared to other national industries, raising therefore the threshold of resources and the risk propensity required for closing the gap.

There was very limited evidence of this Schumpeterian behaviour in Brazilian industry, where import substitution had established reliance on imported technology as a norm. The main exceptions to this norm came from state enterprises (e.g. in the oil, telecoms and aeronautics sectors), where there was a politically driven motivation towards technological autonomy and, more recently, from export-oriented private firms. It was no accident that COBRA, the state enterprise, was a leader in technology investments in the computer industry, but it was an outstanding achievement of BIP that the private informatics enterprises in the mid-1980s were investing between 8 and 10 per cent of their sales on technological activities, in a country where, on average, manufacturing industries spent less than 1 per cent of their sales on such activities. Another noteworthy fact is that national informatics firms allotted about a quarter of their graduate employees to R&D activities, in contrast with their multinational competitors, where less than 6 per cent was allocated (Hewitt 1992).

In order to understand the actual behaviour of the privately owned electronics enterprises it is worth examining in more detail the effects of BIP and other policies upon their investment rationale.

The underdevelopment of the Brazilian science and technology system meant that local firms had to internalize costs and run risks that, elsewhere, were not privately borne, increasing investment risks. The timing and intensity of policy instruments were critical. During the first half of the 1980s, when the scope of BIP was broadened by SEI, public funding of science and technology declined sharply, cutting short the process of structuring the S&T electronics system started during the 1970s. Although this was partially reversed by the Ministry of Science and Technology during its short life-span, when electronics was a priority area, it was not possible to create a critical mass of human and institutional resources. In fact, even top-priority projects, such as CTI, were not properly funded. This lack of resources reflected not only budgetary constraints but also the lack of commitment to the policy from the economic ministries.[2]

The dearth of externalities was not restricted to the top end of technical skills. Although subsidiaries of multinational firms had been operating in the country for decades, staffed mainly with local personnel, the skills they developed concentrated on marketing mainframes, by and large inappropriate to the problems faced by local enterprises striving to design, produce and market different types of products.

The same limitation applied to the network of suppliers of parts and components, which was very rudimentary, since the subsidiaries were supplied mainly by imports. As already mentioned, the consumer durables industry, which in other countries provided a strong market for locally produced parts and components, was supplied mainly by imports throughout the duration of BIP. As a consequence, although import allowances of CAPRE and SEI privileged parts and components, local firms were obliged to verticalize production, spreading their investments and increasing their costs. A side-effect was to encourage the profitability of smuggling parts and components, enhanced by the lack of effective repression. In fact, the main deterrent to smuggling was the moral suasion of the informatics community, which, as previously mentioned, tended to decline with time.

The same tendency towards importing, smuggling if necessary, applied to software, where import controls were both cumbersome and ineffective and incentives to local production were scarce. For microcomponents and software, the international standardization of products during the 1980s led to an increasing pressure to import – legally or otherwise. Consequently, the attractiveness of local investment in such activities, the core of the electronics complex, was reduced.

In other words, in contrast with other countries, the development of the Brazilian electronics complex was plagued by a vicious circle of a cumulative nature, in which the industries which made up the shafts of the fan (e.g. computers, industrial automation) did not benefit from the economies of scale and scope arising from the hub of the fan (components and software). Such negative synergy was internalized by the enterprises which, following the industrial complex approach, undertook several industries and were obliged to follow different and contradictory strategies in their product divisions or subsidiaries.

Added to these external factors is the fact that the local enterprises entering the electronics area had no previous experience in the field, suffering thus all the pains and errors of learning. Although most of them had technology licensing agreements, the knowledge imparted by the licensors was limited, restricted mainly to the provision of product designs and specifications. In most cases, investing, manufacturing and selling was an on-the-job learning process with all its inherent mistakes.

To counter the combined effect of these internal and external factors, BIP had practically only one instrument – the control of imports. Fiscal incentives were almost negligible and the funding from public development banks was very limited, especially during the start-up phase of the industry, when it was most necessary. Moreover such funding was provided under the form of loans and not as risk capital. The undercapitalization of COBRA, where the technology ethos was strongest, was often remarked upon but never corrected by its owners, the state banks. State purchasing power, always present in BIP's documents, as a main instrument for fostering the industry, was used haphazardly and discontinuously, depending on whims and political orientation of the managers of state agencies. Such weakness in terms of instruments and time horizon was not accidental – it reflected the opposition BIP suffered within the state, which, in turn, reflected the conflicts it evoked in the Brazilian society.

The paucity of instruments used by BIP stands in stark contrast with the wide array of policy instruments used in other countries, which ranged from state-intensive and extensive funding of R&D and production to state procurement and support in the internal market and abroad. In such countries, the development of the electronics complex was stimulated by measures which, at the same time, reduced risks and costs. Moreover, in other countries the policies for the segments of the electronics complex were convergent and had a long, usually indeterminate, time horizon, reflecting a wider and more permanent social support.[3]

It is true that the control of imports sheltered the local companies from price and perfomance competition and allowed them to reap considerable margins on the products they sold. If, on the one hand, such margins could be justified on the grounds that, in the absence of alternative sources of funding, they were necessary to allow the firms to make their 'primitive accumulation' to develop local technological and industrial capability; on the other hand, they attracted rent-seekers, who aimed at making the highest possible short-term profits with very limited commitment to the development of local resources. The possibility of smuggling and the limited capability of SEI to control the implementation of the projects it approved (again because of 'budgetary constraints') facilitated the latter behaviour.

In this context, the imposition of a time limit to SEI's power to control imports by the Informatics Law of 1984 probably played a contradictory role: it stimulated both learning and predatory behaviours. However, since the time period was established arbitrarily, as a political compromise between the policy's supporters and opposers in Congress, and bore no relationship to learning curves, its net effect was probably tilted to making the best of the policy while it lasted.

The behaviour of entrepreneurs is obviously dependent on the development of the market they serve. BIP's success was predicated upon the assumption that the Brazilian market would provide a basis wide enough to develop the industry. The increasing role played by static scale economies in the more standardized industries was clearly underestimated by policy-makers. As suggested above, this may be partly explained by the technological emphasis on product design. However, political economy factors may have played a role as well, albeit sometimes unconsciously: if scale economies were prime criteria for guiding SEI's decisions to allow products to be locally manufactured, the agency would be led to restrict the number of local suppliers. This would bring about an inner circle hostility, from the frustrated newcomers, and more strident accusations from policy opposers that it was nurturing an exclusive set of enterprises. As it was, SEI could truthfully contend that it was fostering competition in the industry.

In fact, a trait that distinguishes BIP from other countries' informatics policies is the absence of 'national champions'. From the outset of the policy, at the time of CAPRE's market reserve decision, policy-makers were unwilling to commit resources to one or a few enterprises, which explains the fate of COBRA, an obvious candidate to be 'a national champion'. As a result, the policy tended to follow a 'product design' approach instead of focusing on enterprises and their economics. Such bias was reflected in the staffing of SEI, with very few economists, and in the procedures of analysis, which prioritized technical aspects.

Finally, it is also probable that the estrangement between SEI and the academic community, brought about by the 'national security' origin of the former and the ruthless way in which it replaced CAPRE (where the relationship between bureaucracy and academia was very close) contributed to reducing SEI's capability properly to assess the technical changes at the international level.[4]

Moreover, because of the extent of investment which was required and the structure of funding the investment, based mainly on retained earnings and very expensive, short-term credit, the internal market had to grow quickly – as the enterprises were in the position of a cyclist who must keep pedalling in order to stand upright. Indeed, until 1987 the Brazilian market grew at spectacular rates, impervious to the crisis which affected the rest of the economy during the 1980s. According to data from SEI, between 1980 and 1986 the earnings of national informatics firms grew 7.4 times – about 40 per cent yearly average. In 1987 the growth was 14 per cent – a healthy rate for any industry. None the less, as an indication of the financial fragility of the industry, this decline in growth provoked a serious crisis.

An export-oriented strategy, eventually developed along selective lines, could have provided some solutions. However, pressures to export were applied only on the foreign subsidiaries, which complied in exchange for greater import flexibility. Although local firms did achieve some export success in specific market niches (e.g. services automation), the strategy was very inward looking. Only at its end, in the late 1980s, did BIP begin to consider exports as a priority, but without any policy instrument to back it up.

The inwardness of the policy was the result of several, convergent factors. First, the very success of internal sales reduced incentives to export. Second, the policy itself, by emphasizing local design of products and software, reduced their tradability thus diminishing incentives to export. The scant attention paid to production technology and economics, especially to scale economies, as mentioned above, had important negative consequences for the international competitiveness of the products covered by BIP. This was especially apparent after the microcomputer revolution made standardization and scale economies prime requisites to international trade. Finally, BIP was deprived of an export basis familiar to other developing countries, especially Southeast Asia: the consumer durables industry. While in other countries such industry provided the experience of mass production and international trading for exporting other electronic products, in Brazil the consumer durables goods industry produced limited quantities oriented to the internal market, all the while using mainly imported components. Also in this area, the synergy of the Brazilian electronics complex was largely negative.

The year 1987 was a watershed for BIP. Coinciding with a market slackening was the withdrawal of government support, as shown above. The two events were closely connected. The diminished demand growth of 1987 was largely a result of the failure of the stabilization attempt the year before (the Cruzado Plan), which led the Brazilian government to declare a moratorium on external debt payments and to the weakening of the groups which were the strongest supporters of the policy within the governement coalition. The stepping up of US pressures against BIP, with threats of export retaliations, could not have come at a worst time for BIP. Although BNDES intervened, providing relief credit for some of the largest firms in dire straits, the support of BIP in the economic ministries and the Foreign Affairs Ministry – never enthusiastic – was then substantially reduced, as discussed above. The combination of macroeconomic foreign exchange restrictions with the low priority attached to the success of BIP led to the previously mentioned tighter import restrictions of parts and components, stimulating smuggling and undermining the competitiveness of enterprises that adhered to the policy. At the end of the day, macroeconomic interests reimposed their hegemony over sectoral rationale.

As the policy weakened under internal and external pressures, another cumulative process set in: reading the writing on the wall, many firms reduced their commitment to BIP's objectives and rules and started to negotiate partnerships with foreign companies (using as their main asset their hold on the Brazilian market), increased (legally and illegally) imported content, contained investments in R&D, etc. Such actions undermined BIP further and thus stimulated other firms to follow a similar course, straining the relationships between such enterprises and those that still adhered to the policy. As a result, when the tide of liberalization mounted in 1990 BIP was bereft of support even within the informatics entrepreneurial community, with the exception of a few stalwarts which still upheld the banner of technological autonomy. At present, the latter have adapted to the new norm or are exploring small market niches.

The speed with which a substantial part of the informatics entrepreneurial community changed gears, throwing their lot with a reversal of BIP, may be viewed as a sign of entrepreneurial capability for interpreting market signals and adapting strategies accordingly. Another interpretation is that many of the entrepreneurs were never really committed to BIP's objectives, having supported them as an ideological shield under which they were able to gain rents.

There is probably some truth in both interpretations. Undoubtedly there was a strong component of rent-seeking in the support given by entrepreneurs to the policy. Many only paid lip-service to BIP's technological objectives, using them as a beachhead to enter the market. However, it must be recognized that the limited array of instruments BIP was able to muster and its fixed time horizon were factors, internal to the policy, which stimulated such behaviour or, at least, led the entrepreneurs to keep in mind a policy reversal as a possible alternative. An exclusive commitment to BIP required a Faustian drive to local technological independence for which Brazilian entrepreneurs were never noted. Indeed, one of BIP's achievements was to reveal that, even under unfavourable conditions, there were entrepreneurs capable of such commitment.

A management policy of a technological gap using import controls is based not only on assumptions about the behaviour of entrepreneurs: it relies also on assumptions, explicit or implicit, about consumers' reactions. At the very least it must assume that the latter will bear, with grace or by force, the costs of the policy: the restricted product range offered and the higher price/performance ratio of the goods actually supplied.

For the reasons outlined above, consumers, unequivocally and literally, footed the policy's bill. Since opposers of BIP inevitably pointed out consumer dissatisfaction as a main reason to scrap it, the argument is worth considering in more detail. Schmitz and Hewitt (1992: 31) point out that 'the diffusion of computers made by national firms has been rapid by any standards. The average annual growth rate of the Brazilian microcomputer market between 1984 and 1987 was the highest in the capitalist world at 74 per cent'; showing that the policy, at the very least, did not prevent diffusion. The same authors provide a good survey of the evidence on price differentials between Brazil and other markets at the end of the 1980s, when the attack on the policy was at its peak: they show that the former were double the US prices and not higher than a quarter of the European prices. Over the 1980s, according to the same source, the technological lag in terms of product vintage between Brazilian and international supply had been reduced too. A two-year difference, although declining, was still significant.

No matter how satisfactory the diffusion and learning performances were from a long-term industrial policy point of view, there was a shift from an attitude of co-operation and tolerance with BIP's products to hostility and civil resistance, as expressed by the smuggling of finished goods or by the purchasing of products assembled with acknowledgedly smuggled-in parts and components. Curiously, consumer disaffection increased with the shift from mini to microcomputers, although the purchasers of the former were probably better informed than the latter about price performance characteristics of the international frontier. This is probably attributable to the stronger commitment of the technical profession to the policy.

It is also important to notice that there was no noticeable difference in the performance of national firms and subsidiaries of multinational companies. Moreover, similar differentials of price and product vintages were observable in segments of the electronics complex where BIP was not applicable; for example, consumer goods assembled in the Amazon Free Trade Zone, as well as in other industries structured by traditional import-substitution policies and dominated by multinational firms, such as the automobile and pharmaceuticals industries. Finally, a detailed study of the cost structure of informatics products showed that sourcing was a major cause of the Brazilian higher costs and that import liberalization would not necessarily bring about a reduction of such costs (Schmitz and Hewitt 1992).

In other words, although electronics did cost more in Brazil, it was not significantly different in this respect from the rest of the industry established in the country: its higher costs could not be ascribed to the specificities of BIP. None the less, this was what precisely happened. The press played a major role in this process by carefully reporting all the inefficiencies of the industry and by, equally carefully, abstaining to report its achievements or to compare its performance to other industries and policies. To give an example, it is widely recognized that the automation of the Brazilian banking system is highly successful (Cassiolato 1992). Although millions use it, such success was never linked to BIP.

In this way consumer dissatisfaction, especially of householders, was nurtured, increasing the opposition to the policy. Such opposition, spreading by word of mouth, was probably crucial for its overthrow. The standards may have been double, but they were unequivocally successful in depriving the policy of social support. After all, a well-known apothegm of Brazilian politics is that 'The truth does not count; only versions of it'.

The same may be said about state intervention. In order to pay the external debt other Brazilian industries were subject to import controls similar to those imposed on electronics; however, these were managed by another government agency (CACEX). It was never explained to the public that the import quota of SEI was not decided by itself but by the Treasury. Various industries of the electronics complex, not subject to BIP, depended even more strongly on government incentives, such as the fiscal exemptions of the Amazon Free Trade Zone or the procurement of telecom equipment, which was coupled to technology transfers from the state sectoral research centre. For other industries dominated by multinational companies, such as the electric power equipment, the role played by state intervention was no less important.

We have argued above that BIP's strategists underestimated the problems facing the policy. Such problems could have been reduced by greater selectivity: for instance, by distinguishing between families of products that would have to be imported in the perceivable future, products that could be manufactured locally but with imported technology and, finally, products to which the technology policy of import-substitution could be applied.[5]

As shown above, BIP started selectively, concentrating on small and medium computers; but, soon afterwards driven by the logic of the industrial complex, the scope of the policy was broadened. However, since the policy did not gain control of the complex, the losses were doubled: synergy was negative and the available resources were spread too thinly. By the end of the 1980s the need for greater selectivity was receiving wider recognition – it was a recurrent theme during the preparation of the Second PLANIN. But it was not unanimous: there were those who considered introducing greater selectivity as caving in to the pressures of the US government and internal opposers of BIP.[6] In fact, one of the results of the pressure under which BIP, in general, and SEI, in particular, were placed during the late 1980s was to rigidify the stand of a fraction of the policy supporters, preventing them from making major strategic changes.

Probably, it was too late anyway. As discussed in more detail in the following section, more was at stake with BIP than supply and consumption of electronic products and *delenda BIP* had become a symbol of modernity. Under the prevailing circumstances, no greater selectivity could have saved it.

2.3 Values, interests and social support

A crucial aspect of BIP was its value content. The history of the policy may be seen as the apogee of a *Weltanschauung* and its decline. BIP's paradigmatic force, as a model applicable to other industries, did not escape anyone – supporters and opposers alike, inside the country and abroad. Such attention was obviously reinforced by the strategic role played by electronics in the modern world.

BIP strategists assumed that the notions of national autonomy and local creativity would be strong enough to warrant social support, giving the industry enough time to proceed along learning curves and to approach an international technical and economic efficiency frontier. This support involved the legitimacy of the policy instruments – state intervention discriminating against imports, favouring instead nationally owned enterprises, and local technological efforts and social actors (scientists, engineers, etc.) responsible for such activities.

Although a considerable part of the debate surrounding the policy was conducted using such 'objective' arguments as the performance and price of the products of BIP, the objectiveness was more apparent than real, since what was at stake in the economic and technical efficiency debate was an income distribution question – how costs and benefits of the policy would be distributed along different social groups over an imprecise time duration – a question which cannot be answered without resort to value judgements. Moreover, underneath such arguments lay unquantifiable values about creativity and power: of local versus foreign agents; of producers versus consumers; and of the state versus the market. The fierceness of the debate, which often turned vicious, and the polarization of opinions are indicative of the indivisibility of such values, which rendered impossible a meaningful compromise.

As shown above, the balance of forces shifted with time and it is worthwhile to try to identify the factors conditioning such movement. Let us begin with the contextual factors, since BIP was a sectoral expression of a pattern of development. As mentioned, at the inception of BIP a similar pattern was applied to other sectors. Expansive macroeconomic conditions, a strong state which led the economy and had an assertive international position, and a confident outlook towards the future, supported the assumptions of BIP. Although the macroeconomic conditions worsened considerably during the first years of the 1980s, by 1984 an export-led recovery was under way, lending credence to the view that the crisis had been overcome and that the Brazilian economy was bound to return to its 'natural', high-powered course. Optimism was greatly reinforced by the imminent end of the military regime. The approval of the Informatics Bill by Congress fits into this context.

In turn, the debate on the bill contributed to make informatics a national issue and a symbol of state intervention, focusing the opposition to the latter on BIP. If at its inception the informatics policy was a kind of guerrilla warfare, as aptly described in the literature (Adler 1986; Dantas 1988), its institutionalization by Congress turned the conflict into a war of position, with the occupants of the trenches well defined in ideological terms.

The débâcle of the 1986 Cruzado Stabilization Plan put an end to optimism and the next year ushered in a period in which the unattended Brazilian structural problems claimed their price as an economic crisis still under way. The credibility of the Executive, which was already weakened by successive unsuccessful attempts to reduce public deficits, resulting only in a deterioration of the state apparatus, was finally swept away as the economy bordered hyperinflation. This could not but affect the legitimacy of a policy, such as BIP, which hinged on state intervention.

None the less, the legitimacy of the values underlying BIP was still strong enough to lead Congress to enshrine in the 1988 Constitution a definition of 'national enterprise' closely patterned upon the BIP, a characterization of the Brazilian market as a 'national asset', and to reserve for the state certain strategic sectors such as oil and telecom. However, the same Congress established, against the strenuous efforts of BIP supporters, the maintenance of the privileges of the Amazon Free Trade Zone. In fact, the process of writing up the Constitution led to the consolidation of the conservative forces within Congress and the articles which supported BIP were a compromise with the waning nationalist faction. To all practical purposes, they were the latter's swan song.

The disenchantment of public opinion with state intervention due to internal reasons was further deepened by the international tidal wave of neoliberalism which swept the world during the 1980s. The media, which had been strenuously campaigning against state intervention since the late 1970s rode high on the wave and singled out BIP as a prime target. In this way the conflicts with the US over BIP were well explored.

Much has been made of the US pressure against BIP as a cause of its demise, but its role and strength can be gauged only against a background as described above, in which loomed the huge Brazilian external debt. As shown by the case of the nuclear policy during the second half of the 1970s, under different economic and political conditions even stronger pressures were sustained by Brazilian governments. None the less, under the conditions prevailing at the end of the 1980s, the US pressure played an important role in the reduction of the government's support of BIP, as previously examined.

In short, during the second half of the 1980s the macroeconomic and political conditions became highly unfavourable to BIP's values. The latter were further undermined by the policy's specific problems of managing the technological gap, as previously commented upon. The behaviour of entrepreneurs and consumers shows that the *mores* of the Brazilian society were mismatched with BIP's in at least three critical aspects. First, on the valuation of independence, technological and otherwise. Public opinion, as expressed by consumers, wanted Miami not the Silicon Valley. Entrepreneurs were quite willing to parlay autonomous decisions for a safer existence. Second, the time horizon: although industries take as long as human beings to reach maturity, as shown by other international experiences (e.g. the Japanese car industry), in Brazil there was little patience with BIP. Third, a cavalier approach to the law, as witnessed by smuggling and other corner-cutting procedures. The latter points to a deeper problem, which lies at the heart of BIP – the lack of legitimacy of the state.

As shown, throughout BIP's life-span its supportive coalition remained essentially the same: academics, technicians and bureaucrats – social actors with professions that led them to place high value on independence, especially technological independence. Added to this coalition were the entrepreneurs of the sector, with the misgivings already mentioned. Such an alliance, which had to fight against a growing number of opposers as BIP encroached on a broadening range of interests, could never convince public opinion that it was fighting for the common good. The media said and people believed that it was only a particularist policy – 'for some colonels and a few inefficient entrepreneurs'.[7] It was not hard to believe, given the record of particularist policies of the Brazilian state, strongly reinforced during two decades of military regime. In fact, BIP was often branded as an exclusive product of the military, in spite of being one of the very few sectoral policies submitted to Congress.

The fact that policies alternative to BIP served other particular interests was conveniently disregarded by its opposers. However, the important point is that BIP failed to capture the hearts and minds of the majority of those involved. Therefore, as a result of the general conditions and of its specific problems, at the end BIP was deprived of most social support. Accordingly, it ended with a whimper, not with a bang.

3 WHAT IS LEFT

The metaphor of the half-full glass, which is also half-empty depending on who looks at it, is applicable to BIP as an industrial and technological policy. It produced a sizable industry with considerable technical capabilities, although far short of its ambitions. Its products, in terms of price, performance and uptodateness, are similar to those of industries spawned by the traditional import-substitution model. It provided jobs which involved the use and development of technical skills which other industries in the country did not foster but which they now can use. Moreover, it showed that under appropriate circumstances there are some Schumpeterian entrepreneurs in Brazil.

However, as an ideological construct – as the proof that a more independent pattern of development was both desirable and feasible – in pragmatic terms, BIP failed. In the end it had very little social support, no matter if justified or not. As pointed out above, several lessons about the objectives and *mores* of the Brazilian society can be drawn from such experience, as well as about the relative weight of interests and how they operate, within and outside the state. As for the latter, it is a reminder of the limits and penalties attached to *hubris*.

At the same time, looking again at the glass, BIP was an example of audacity, hope, hard work and honesty – qualities that are not amiss in a society such as Brazil. Moreover, since the values that inspired it are not dead, they may find a different form of expression, eventually as a new policy. Opposers of the policy may read this as a threat and think of Dracula but its supporters may find relief and hope in the image of the Phoenix.

NOTES

1 I thank Arthur Pereira Nunes, Ivan da Costa Marques and José Guaranys for the time they spent discussing and recollecting facts and ideas about the policy. The final version of the paper benefited from the careful reading by Wilson Suzigan of its first version, as well as from the comments of the participants of the seminar on The Politics of Science and Technology Institutions in Latin America, held at INTECH in April 1993. The usual disclaimer that the article represents strictly personal views obviously applies, but, since the policy was laden with values, as argued in more detail in the text, it is convenient from the beginning to 'declare an interest': in my double capacity, as a bureaucrat and academic, I was an active supporter of BIP. Since 'objectivity' is not to be confused with 'neutrality' I hope I have provided a balanced account of the policy but corrections are all the more welcome.

2 Even if funding had been appropriate, the maturation period of investments in S&T would probably have been out of synchronization with industrial needs. To counter this, the Ministry of Science and Technology established in 1987 a

special scholarship programme for the high-tech industries, aimed at training the personnel of enterprises (RHAE – Recursos Humanos para Areas Estratégica – Human Resources for Strategic Areas).

3 For an account of policies of developed countries, see Jowell and Rothwell (1986). For a comparison between BIP and policies followed by Japan, India and South Korea see Evans (1992).

4 This point was forcefully brought to my attention by Ivan Marques during our conversations.

5 This was the essence of a proposal I put forward to SEI in 1988 and at a seminar held at the University of Campinas to evaluate BIP in 1989 (Erber 1989). See next note.

6 A personal anecdote may illustrate this: after I gave the presentation at SEI (see note 5), one of its officers burst into my office (I was Assistant Secretary-General of the Ministry of Science and Technology), accusing me of 'playing the Americans' game', since they had required selectivity in the policy too. For the record, such a hard line was not followed by SEI's Executive Secretary, who fully agreed with the need for more selectivity.

7 The unblemished record of SEI in terms of corruption testifies the strong *esprit de corps* and the sound ethos which pervaded the agency. Most of its officers could have earned more elsewhere.

REFERENCES

Adler, E. (1986) 'Ideological guerrillas and the quest for technological autonomy: development of a domestic computer industry in Brazil', *International Organization*, vol. 40, no. 3.

Bastos, M. I. (1992) 'State policies and private interests: the struggle over information technology policy in Brazil' in H. Schmitz and J. Cassiolato (eds) *Hi-tech for Industrial Development: Lessons from the Brazilian Experience in Electronics and Automation*, London, Routledge.

Cassiolato, J. (1992) 'The user-producer connection in high-tech: a case-study of banking automation in Brazil' in H. Schmitz and J. Cassiolato (eds) *Hi-tech for Industrial Development*, London, Routledge.

Dantas, V. (1988) *Guerrilha tecnológica: a verdadeira história da política nacional de informática*, Rio de Janeiro, Livros Técnicos e Científicos Editora.

Erber. F. (1989) 'Politica industrial nacional e política de informática nacional', Instituto de Economia, UNICAMP, mimeo.

Evans, P. (1992) 'A informática no Brasil, India e Coréia na década de oitenta: uma análise comparativa da política e da organização industrial' in P. Evans, C. Frischtak and P. Tigre (eds) *Informática Brasileira em Transição: Política Governamental e Tendências Internacionais nos Anos 80*, Rio de Janeiro, UFRJ/IEI.

Gaio, F. (1992) 'Software strategies for developing countries: lessons from the international and Brazilian experience' in H. Schmitz and J. Cassiolato (eds) *Hi-tech for Industrial Development*, London, Routledge.

Hewitt, T. (1992) – 'Employment and skills in the Brazilian electronics industry' in H. Schmitz and J. Cassiolato (eds) *Hi-tech for Industrial Development*, London, Routledge.

Jowell, P. and Rothwell, M. (1986) *The Economics of Information Technology*, London, Macmillan.

Piragibe, C. (1985) *Indústria de informática: desenvolvimento brasileiro e mundial*, Rio de Janeiro, Editora Campus.

Schmitz, H. and Hewitt, T. (1992) – 'An assessment of the market reserve for the Brazilian computer industry' in H. Schmitz and J. Cassiolato (eds) *Hi-tech for Industrial Development*, London, Routledge.

Part III

Conclusion

7 Politics, the state and policies for science and technology in Latin America

Maria Inês Bastos and Charles Cooper

> Only pragmatic experiments can show when and where intervention is the lesser evil. For we have to act in this least bad of all feasible worlds, in which everything is for the *n*th best. The challenge consists in designing institutions that combine an appeal to private initiative and enterprise with social objectives and public accountability.
>
> (Streeten 1993)

The previous chapters have dealt with two major questions. One is what capacity the Argentine, Brazilian and Mexican states have exhibited to date for the promotion of science and technology activities. The other is how far the existence or lack of such a capacity is to be explained by political factors. These two questions demonstrate the scope of the study. The objective has been to explore political conditions for science and technology policy design and implementation, not to describe the outcomes of public policies for S&T, nor to explain state activity in this field.

In this concluding chapter we contend that we need to understand the role of political institutions, political constituencies, forms of political representation, and political culture as factors in the microeconomic decision-making process as a necessary complement to the economic approach which has been dominant in technology policy studies in Latin America. Strategies and modalities of technology investment decisions cannot be understood exclusively as economic phenomena. Political-economy analyses of S&T policy in Latin America have brought forward many aspects which are hidden by a purely economic approach to technology development and innovation. Some additional issues are presented in this volume. Hopefully, others only suggested here will be uncovered in future research.

The authors have used different theoretical frameworks to deal with these questions. We think this has enriched the analytical content. Authors have also analysed with different degrees of detail the relationship between state and society, the role of ideology, implications of political regime, the constraints and advantages stemming from the international system and market, and the political dynamic among and within S&T organizations of the state apparatus. Nevertheless all chapters offer an analysis of central aspects of a political sociology of S&T policy in the most developed industrial economies in Latin America; analyses which make it possible to draw some conclusions about common features and trends. This is what the present chapter will do.

Summaries of main findings about state capacity for building national systems of innovation in Latin America and about political factors that shape this capacity are presented, respectively, in Section 1 and Section 2. The final section discusses prospects for state activity in support of science and technology in Latin America. There are two issues here. First there is the question of whether the state has an active role in science and technology policy in open economies which now characterize Latin America. The other is that circumstances of economic crisis and reform have in themselves produced a situation in which the conceivable role of the state is inherently more constrained than it used to be.

1 GOVERNMENT'S CAPACITY FOR BUILDING NATIONAL SYSTEMS FOR THE SUPPORT OF INDUSTRIAL INNOVATION IN LATIN AMERICA

A main conclusion from the contributions to this volume is that the governments of Latin America's largest countries have shown strong capacity in support of innovation only through a few of their various agencies and in very specific industries. In these areas the states in the region have been relatively active either in reinforcing the private process of accumulation or in responding to strategic considerations. Nun (Chapter 2) discusses the cases of efficient agencies in the area of nuclear power and in the promotion of innovation in agriculture. Bastos (Chapter 3) mentions sectoral pockets of efficiency in petrochemical, defence, communication and the computer industry. Nadal (Chapter 4) refers to successful activity in the areas of health, agriculture and energy. Beyond these sectoral experiences, it is generally accepted by the contributors to this volume that the level of science and technology activities by private firms is deficient, that systemic linkages between research institutions and industry are still lacking, and that local research is of diverse and generally not very high quality. As many decades of state policies have failed to correct or significantly to change this situation, it can be concluded that the state has shown limited capacity for building innovation capability in Latin American industry. Thus the contributing authors concur explicitly or implicitly with the view that 'national systems of innovation' (which have received much attention in the recent literature on technology policy) have developed only to a limited extent in the countries under study – indeed perhaps not at all. Bastos and Erber argue that there has been some such development in Brazil; Nun and Nochteff are much less optimistic about Argentina – as is Nadal about Mexico. This pessimism is both understandable and, probably, overdrawn.

One reason for the exaggerated depiction may devolve from the powerful influence of ideas about 'technology dependence', which for many years led to a lack of appreciation of the actual technological developments in Latin American industries. As Katz (1976) and his colleagues pointed out early in the debates about *dependencia tecnologica* (with empirical evidence to support their claims), there has been local innovation in Latin American industries and often a significant development of local capabilities, at least during the period of protected economy. However, not only has local innovation received limited recognition, but at the same time, these less noticeable successes were often overshadowed by the relative failure of technology policies at the 'grand' level. There were many less successful efforts to develop high-technology industries, and to encourage domestic generation and adoption of sophisticated production technologies. The failure to recognize modest successes along with an understandable preoccupation with large failures produced pessimism, but the successes were nevertheless there.

Another source of pessimism is the contrast between the state's grand rhetoric in designing ambitious technology policies and its modest capacities to implement them. Evidence on this last point, while found everywhere in the region, was manifest in Argentina and Brazil more than in Mexico where the state has shown higher stability and policy consistency. Clearly this is easier to recognize in hindsight, and it is perhaps inherent in the politics of development policies, but it is plainly a potent source of disillusionment. A constant flow of communication between those who design the policies and those who will have to make them work is, however, a fundamental precaution that could reduce the probability of defining objectives too grandiose in the face of available resources. The implications of 'embeddedness' of technology policies in Latin American society were explored by Nun (Chapter 2), Bastos (Chapter 3) and Nadal (Chapter 4) and will be summarized in the next section of this chapter.

A third source of pessimism with regard to state capacity for inducing technology activity in Latin America is related to the concept of science and technology policy itself. For more than two decades, policies for the support of basic research, R&D, technology transfer and technology change, particularly in industry, have been referred to as the composition of one single set of policies, which were called S&T policies. Evidence suggests that this is a misnomer and that there is a sharp distinction between the different areas of policy which, by convention, have been lumped together. The contributions to this volume show that much of what was described as S&T policy, in fact amounted to support for academic research. It was essentially about science policy. The technology component was handled in other institutional settings and the S and T system was frequently judged negatively for not dealing with it. It does not follow that technology policy was somehow overlooked.

Bastos (Chapter 3), Nadal (Chapter 4), and Nochteff (Chapter 5) note that S&T policy institutions were never given the power they would have been entitled to had the governments in the region a more than verbal commitment to make them effective in the promotion of technology development. This is true, but can also be interpreted the other way round: that the organizations formally in charge of S&T policy have never had, in fact, other attributions than to administer funds, for graduate training and research activities at universities and for other publicly supported research institutions. In this respect they continued, from the mid-1970s when they were created or transformed to formally include technology development considerations, with the same activities they (or their 'parent' agency) had been doing for decades. These activities are generally accepted as part of what is called 'science policy'. CONACYT, CONICET and CNPq are science policy agencies and probably should continue to be so. If they manage to stimulate a closer relationship between university and industry, this may prove to be not only financially useful but, mostly, 'scientifically' fertile.

The important consequence of identifying what these institutions have really been doing under the label 'science policy' is to draw attention to the other aspects of this policy they have not been able to tackle. One such aspect, and a central one, is the inadequate quality of the national scientific base which points to insufficient links between science policy and education policy. The latter not only has failed to provide the region with the basic educational and vocational skills necessary, but also has not given enough support to improve the quality of secondary and tertiary education.

Technology policy in the region has definitely been carried out through economic and industrial policies where issues of trade policy, intellectual property rights, and financial incentives, to mention only a few, have been settled without more than formal consultation with S&T policy agencies. Organizations of what Bastos called the 'planning cluster' such as BNDES and FINEP in Brazil, the ROTT of Mexico analysed by Nadal, and Argentina's CNEA or even INTI, are in practice the agencies which carry out technology policies, even if this is not always intentional. Their concern with innovation or international competitiveness and eventual utilization of domestic engineers and research has not been substantially affected by the constitution of a formal S&T system in the region. Given this institutional background, it is not surprising that the concepts of implicit and explicit technology policy have had such an important role in Latin American policy discussions.

Public support for basic research, particularly in universities and publicly funded research institutes, and for graduate training within and outside the region, has absorbed most of the public budget earmarked for S&T activities in Latin America. Its effect on technology development has been very scant in the region, as anywhere else. In fact, the contribution of national capabilities in basic research to technological innovation is indirect and definitively not primarily based on the research content in itself (Mowery 1994). In Latin America, the strongest such contribution of basic research has been in the training of scientists and engineers who constitute significant resources eventually mobilized for technology adaptation. Some portions of university research, mainly from the few 'centres of excellence' which had resulted from decades of science policy and are closer to regional industrial centres, found direct applications particularly when reduction of public funds pushed the universities to a systematic search for partnership with industry (Chapter 3). The political by-product of state support for basic research and graduate training in the context of insufficient demand for such highly trained researchers and engineers has been the strengthening of a 'scientific constituency' whose survival is mostly dependent on continued public support. State agencies in charge of administering public funds for research have had to withstand strong pressures from this constituency, particularly when an aggravated economic crisis and stabilization policies reduced the available resources. In the largest countries such as Brazil, Mexico and Argentina, these agencies have for a long time lived with public pressure to alter the national scientific research agendas in order to connect publicly funded basic research with technogical needs. This has met with strong opposition from their scientific constituencies. Given their role, these agencies have, in fact, evolved by reflecting the interests of the scientific communities. It is not always easy for them to give primacy to concerns about the consistency of S&T policy with national development objectives.

The most significant state support for technology innovation in Latin American industry was definitely embedded in the import-substituting industrialization. These policies mixed import and adaptation of foreign technology with investment in state-controlled research institutes and a general protection against foreign competition. The high growth rates of the mid- and late 1960s were accompanied by the highest levels of industrial growth in the history of the largest countries in the region together with a sustained rise in productivity and in the share of manufactured goods in total exports. While the main source of technology remained foreign, the technological effort in adaptation contributed to a gradual development in some sectors and consolidation in others of an internal technological capacity of the manufacturing industry. Technology and engineering services of local origin began to be exported to other countries and the development of more sophisticated industries was initiated. This process was interrupted in these countries, albeit with different timing, by the various attempts at solving the economic problems that surfaced with the oil crisis and the subsequent debt crisis.

As far as the discussion in this book is concerned, the main results of policies of import substitution were not only the creation of a relatively integrated industrial sector, particularly in the large countries, and some internal technological capacity, but also the crystallization everywhere of strong protectionist interests and some disincentives to innovation. Protectionist policies went through a process of what Nun (in Chapter 2) calls 'naturalization' by which they entered the prevailing culture of the economic agents, domestic and foreign. Explicit technology policies seeking to change this were perceived as political interference and were resisted accordingly by relevant economic agents. It was only recently, with the emergence of an acute economic crisis in the region and after one decade of unsuccessful attempts at dealing with foreign debt and macroeconomic instability without forgoing the culture of import substitution, that domestic conditions grew ripe for a strategy of reduced protectionism and open markets. This strategy centres around the push for increased international competitiveness. This, in turn, may suggest that technology change has a new primacy in policy. As we shall later show, this does not always follow.

2 POLITICAL AND ADMINISTRATIVE CONSTRAINTS ON STATE CAPACITY FOR PROMOTING INNOVATION CAPABILITY IN LATIN AMERICA

What are the major political and administrative factors that were found to help explain the limited state capacity in Latin America for supporting innovation in industry? The authors find political reasons in the peculiar relationship the state has had with society. This has resulted, on one hand, in the lack of, or uncertain, social support to policy initiatives and, on the other, in the lack of, or very slow reactions by governments to, changes in technology and society. Characteristics of the policies themselves were also found to be internally conflictive and this contributed to the erosion of their social support. In addition to its failures in generating and sustaining general political support for its technology policies, the state was also found unable to get the decisive co-operation of industrialists, and to a lesser extent of domestic engineers and researchers, for its more specific projects on technology development. Inadequate political appeal, inflexibility and a lack of adaptability of technology policies were also found to be linked to the state apparatus corporate culture which structured much of the voluntary technology initiative of the state around a project of 'autonomous technology development'. These findings are summarized here. They are organized in three sub-sections.

2.1 State–society relations and technology policies in Latin America

Contributors to this volume discuss two aspects of state–society relations that have negatively affected technology policies in Latin America. One refers to the relationship between the state and technology policy constituencies and the other to the effects of state autonomy and insulation from society on technology policy implementation.

The relationship between the agencies in charge of technology policy and two of the main actors of technical change and central components of the policy's constituency, industrialists and engineers/researchers, has been deficient or even conflictive in the large countries of Latin America. This is clearly demonstrated in the contributions of the present volume, as discussed below.

Many of the technology policy goals, as 'technology self-determination' of the Brazilian and Argentine informatics policy, did not emerge from the real economic process of accumulation and competition but were exogenously imposed by the state. Consequently, industrialists have shown a systematic lack of interest in taking technological considerations into account in their investment decisions, or have not maintained their commitments to technology autonomy when market signals changed or the policy lost major political support. Erber in particular shows this lack of consistent commitment. Lack of interest or withdrawal of support of the large Argentine economic groups is considered by Nochteff as probably the most important factor in deciding the relative failure and eventual abandonment of the electronics policy. Beyond the sectoral level, he also argues that liberalization initiated in the mid-1970s in Argentina produced a reversal of the infant industry argument in the sense that sectors which had previously been considered worth protecting because of technological learning processes they might engender were opened to international competition and were not able to survive. Liberalization also displaced technological considerations from the agendas of economic agents as well as the Argentine government. This is surely true of other states, although not specifically tackled by the other authors.

Very much present in the discussion about the relationship between state and industrialists in matters of technology policy (as pointed out in the previous chapters) is the notion of a shortage of entrepreneurial skills among Latin American manufacturers. Latin American entrepreneurship is variously criticized by the contributors. Nadal points to the inclination to seek high profitability aimed at short-term recovery of investment; Nun to an unwillingness to take the risks of competition in trade; Erber describes a willingness to exchange autonomy of decision on technology in return for a safer existence; and Nochteff mentions a technology behaviour basically adaptive and not seeking the frontier of 'best practice'. These views seem plausible and, albeit lacking detailed empirical support, they point to a central issue for technology policy in the region. These contributions suggest that in Brazil and Argentina the state has not developed a capacity to temper the operation of market forces and to convince industrialists to do what it considered necessary and desirable, a capacity which some Southeast Asian states are sometimes presumed to possess. On the other hand, there has not been an effective way for the industrialists to convince the state officials of the foolishness of policy decisions or the illusory nature of their technology development goals. The nonexistence of two-way channels of communication between agents of technology change and governments in Latin America seems to have been one of the fundamental reasons for the impaired efficiency of technology policies in the region.

In relation to researchers, restricted demand for their activities has reinforced their stereotypical attitude of disinterest in the practical applications of knowledge. In some cases, the lack of interest in practical applications of knowledge on the one hand, and political engagement on the other, produced what Nun calls a 'defensive withdrawal' of Argentine researchers which has further deepened the gulf between the realities of production and domestic achievements in research and development. In the other two countries, tension between researchers and the state has had two other dimensions.

One is mostly of an ideological nature, and during the period of military rule it transformed into open conflict. Political repression of dissidents in that period did not spare universities and research institutes in Brazil (Chapter 3) or in Argentina. While in Brazil it was coupled with initiation of graduate programmes in various fields and the provision of resources for research, in Argentina it was accompanied by a drought of research funds.

The second dimension emanates from the collapse of the financing capacity of the state and the more recent dominance of orthodox stabilization strategy. Stabilization measures have resulted in a drastic reduction of resources for universities and public research institutes, the threat of closing down research units of S&T policy agencies and the disbanding of the group of S&T policy managers. Across-the-board reduction of public investment in S&T activities both at universities and public research institutes jeopardized the effort that had been made during decades of S&T policy. The survival of the technologically active research teams within public institutes (e.g. CPqD of Telebras in Brazil, and CNEA in Argentina) has been also threatened by the process of privatization. In the case of Mexico, economic neoliberalism has had another effect upon university research staff: conflict has emerged around the policy of keeping researchers' wages so low that they are pushed to other occupations for supplementary wages, including scholarships provided by CONACYT. It is doubtful whether neoliberal concerns with international competitiveness include consideration of technological change. If they do, they will have to face the contradictory implications of destruction of scientific and technological research capability in the interests of fiscal equilibrium.

Technology policies in Latin America have emerged mostly as an autonomous action of the state. While Bastos argues that the emergence of a project of technology development in the late 1960s in Brazil was possible as part of the process of industrialization, it was nevertheless put forward by the administration, to a considerable extent unnecessarily, as a component of an autarchic strategy of development. Initiatives in Argentina and also, to some extent in Mexico, followed this model and were justified on similar grounds.

The authors see some relationship between the state's autonomous initiative in technology policy and the policy's modest results. Bastos and Nadal explore the implications of the political regime on predispositions towards taking policy initiatives and towards their effective implementation. For Bastos, the insulation of the authoritarian state, which gave it almost absolute freedom to design policies with grandiose technology goals, also limited the effectiveness of policy implementation. Nadal pointed out that the operation of technology policy instruments in Mexico has relied on high discretionary powers of officials in charge, and that this has made them prone to manipulation by interest groups. Bastos's findings are supported by recent studies about the effects of executive authority on initiation and consolidation of economic reform in Latin America. The consolidation of both the constitutional system and the economic policies in the region were found to be ultimately dependent 'on the development of institutionalized channels of representation which provide the basis of political support for a given policy regime' (Haggard and Kaufman 1994).

The relationship between political regime and state efficiency in policy implementation is still obscure. The direction of causality is unclear and the available evidence is inconclusive (Bardham 1993; Huber, Rueschemeyer and Stephens 1993; Przeworski and Limongi 1993; Sørensen 1993; Lijphart and Crepaz 1991). However, because of the experience of developing countries that were able to implement efficient technology (and economic) policies, the general belief seems to be that authoritarian regimes fare better than democratic ones.

Have the present political conditions now become more adverse for successful technology policies in Latin America in view of the fact that everywhere in the region, with the possible exception of Mexico, authoritarian rule has already disappeared? Probably not. The contributions to this book contend that authoritarianism is not an asset but a liability for the implementation of realistic technology policies, and that the restoration of constitutional order creates a more conducive environment for such policies. The rationale for this conclusion is found in Nadal and Bastos. Nadal argues that the nonexistence of a democratic government in Mexico not only removes from the public eye the possible manipulation of policy design and implementation by interest groups, but also fails to provide institutional channels for articulating legitimate interests whose concurrence might be needed for the implementation of technology policy. In addition, he found that the lack of an appropriate juridical institutional framework keeps public authorities mostly unaccountable to civil society. This insulation and unaccountability, instead of generating favourable conditions for state efficiency, fuels policy rigidity and increases the chances of capture by private groups. Bastos argues that sustained change of the traditional risk-aversion behaviour of industrialists into a more innovation-oriented behaviour – a major goal of technology policies – is better accomplished via compliance and co-operation. When industrialists are involved in the process of policy design and there are open channels for them to circulate their demands and complaints, then it becomes more likely that policy goals and industrialists' actions will converge. This convergence through negotiation and consensus-building, which has been argued to be at the centre of East Asian success, also seems to help explain the relative success of some sectoral technology policies in Latin America. In addition, signs of a less sectoral and more generalized participatory attitude expressing a healthy initiative on the part of civil society were already identified, at least in Brazil (Reis Velloso 1994). This conclusion makes a case for a change in the style of management of technology policy in the region; which may be easier to achieve in the context of constitutional order and increased accountability than it was under authoritarian rule.

The conclusion reached recently in relation to economic policy reform in Latin America could also apply to technology policies in the region: 'democratic politics – redefining the relationship between economic policy-makers and the polity from insulation to engagement – is vital to the success of the economic reform process, rather than a factor in conflict with it' (Bradford 1994: 21). Essential to this process is the redefinition of the role of the state in Latin America with emphasis on the need to build or strengthen its abilities to articulate and empower other economic and social agents. This is a big change for which the present states in Latin America may not be well equipped. Much of what has to be done is dependent on the relationship the state can establish with society: accountability and transparency will have to be enhanced. These developments will affect the general capacity of the states for policy implementation in all areas. Specifically in relation to technology policy, a new culture will have to be developed in which collaboration with economic agents is seen as something to be sought and not avoided.

2.2 Administrative capacity, corporate culture and effectiveness in technology policy

If there was no strong demand from within the economies and societies, what pushed autonomous states to initiate S&T policies in Latin America? The influence of international actors and programmes, particularly UNESCO's and 'The Alliance for Progress', has been discussed elsewhere. It is raised again by Nun in Chapter 2 and Bastos in Chapter 3. Obviously, domestic political and economic considerations may predominate. In the Brazilian case, the initiative for science and technology policies was justified by the broader strategy, of the developmental and nationalist state, of building legitimation on the basis of the substitution of 'technical knowledge' for 'political' considerations as central criteria of decision in all spheres of government.

There is a further point of more general applicability about the establishment of technology policies in Latin America which can be drawn from the discussions in this volume. This relates to some of the influence of the anti-dependency culture, which in some cases ultimately found an expression in the technology policy agencies.

It is generally agreed that technology policy concerns did not come on to the agenda of Latin American governments until the late 1960s or early 1970s. When they appeared they reflected an incipient contradiction between a growing antidependency concern (and associated preoccupation with a need for technological self-reliance) on the one hand, and, on the other, all the interests which had grown up around protected national economic systems which had become importantly dependent on foreign capital and production technology. For some the need for technological self-reliance became a key policy ultimately necessary to reduce the role of foreign capital within the economy. For others – elements of local industrial classes, for example – such policies seemed potentially useful because they held out promise of state support in the purchase of foreign technology and negotiation with foreign enterprise. For yet others, the technology policies threatened the alliances between the state, local business and international firms, which underlay their whole political programmes. The emergence of technology policy concerns in the early period in Latin America reflected the uneven way these political interests worked themselves out in various countries. In Brazil and Argentina, for example, the state is something of a divided Leviathan and its component parts have enough room to manoeuvre, to act independently and disconnected from each other. In these countries nationalist technology policies found political and administrative space to flourish. They successfully resisted the more internationalist, foreign capital and technology-oriented modes of political thought for substantial periods, before finally being absorbed. In Mexico, the state apparatus seems to be much more integrated with little room for divergence in orientation among its agencies. In Argentina and Brazil the final expression of the antidependency, nationalist technology policy thrust may be found in the informatics sector policies. A Mexican equivalent of the Brazilian and Argentine informatics policies was very short-lived, though technological self-reliance ideas encountered considerable credence in Mexico at somewhat less dramatic levels – for example, in the quite complex attempts to control the inflow of foreign production technologies, more generally in the manufacturing sector.

When the concern with technology development drove the governments in the region to define explicit technology policies, this task was conferred to the existing organizations of the public administration which had been supporting basic research. Technology development was appended to their dominant activity, with the consequent poor results already mentioned. As we have seen, the reason for these was often that the bulk of technology policy was effectively decided and implemented outside the organizations explicitly in charge of S&T policy and very often in different directions.

One response was to seek the centralization of such policy-making activities in an agency placed high in the state hierarchy. Another was the creation of organizations close to the Presidency with advisory functions.

The clearest such attempt in Latin America was probably the creation in 1985 of the Ministry of Science and Technology (MOST) in Brazil, which emerged within the administrative reform of the new civilian administration. The Brazilian MOST was also a political response to demands for a stronger commitment to scientific and technological development made by the scientific community who closely monitored the activities of the new ministry (Chapter 3). MOST was never institutionally strong and did not do much more than earmark a larger slice of the federal budget for basic and technology research. In the subsequent crisis of institutional instability of the Sarney, Collor and Franco administrations, MOST was successively demoted, extinguished and restored.

There were similar evolutions elsewhere. In Argentina, SECYT was created in 1984 to advise the President in all matters relating to S&T, to design policies for S&T development, and to promote research, financing and transfer of know-how. In Mexico, there is no such centralized agency in charge of science and technology development. CONACYT, created in 1971 by transformation of Mexico's National Institute for Scientific Research (INIC), received the attribution of advising the agencies of the public sector on matters of R&D, technology transfer and training of human resources in addition to its executive powers to provide additional support to science and technology organizations. Within a line of science policy which, according to Nadal, has more symbolic than practical motivations, the Consultative Council for the Sciences (CCS) was created in Mexico in 1989. It functions as an advisory body to the President's office and channels all contributions of the scientific community to the national development plans.

Attempts at centralizing the definitions of broader strategies have systematically failed because of an inability to adjust ambitious objectives and scarce political and material resources and difficulties to relate to the sphere of production. As most of these centralizing experiences were launched in a period of economic stagnation and restrictions in public spending, it is not clear whether these co-ordinating agencies are redundant or not: that is, whether their existence would make any difference at all for the improvement of technological activities in the region; or whether their lack of success merely reflects the collapse of funding. Presently, neither SECYT in Argentina nor MOST in Brazil seems to have found its proper position in the implementation of S&T policies; each has had to compete for resources and attributions respectively with the subordinate organizations CONICET and CNPq. The effectiveness of those central agencies has always been impaired either by lack of institutional, political and material resources, or by what Nun called 'the recurrent lack of adjustment between always ambitious objectives and always scarce resources'. Realigning with its younger brother institutions CONICET in Argentina and CONACYT in Mexico, the Brazilian CNPq has recently returned to its old activities of using the scarce resources available to support academic research.

Technological self-reliance was the main goal of explicit S&T policies in Latin America and corresponded to a critical political aspiration of those segments of society at large which sought to find a path to self-sustained, less vulnerable and independent economic development. That goal was consistent with an autarchic project of import substitution based on the belief that it was desirable and possible to endogenize successively all fundamental sectors of the economy and domestically to create an autonomous process of technological progress (Frischtak 1993). Industrial autonomy rather than gains in productivity constituted the fundamental principle of S&T policy. This was complemented, in some specific cases such as the Brazilian informatics policy, by privileged support for the locally owned industry. In this latter case, the political principle of support to locally owned companies was also based on the fact that there had not been a transfer of technology innovation capabilities to local companies involved in joint ventures with international companies (Chapter 6).

The autarchic project resounded strongly within the Latin American state technology bureaucracy, whose approach sometimes evidenced an active enthusiastic belief in their power to change reality, or a more passive, disenchanted vision of what was possible. The Brazilian information technology policy may be seen as a product of just such an enthusiastic response by the technology bureaucracy to an autarchically inclined project. The project was not strictly autarchic as it was always accepted that technology imports would have to be a necessary part of the technology strategy. There was, however, a strong assumption that, with time, they could be replaced by local alternatives. Scientists and education administrators within the state science policy bureaucracy represented a different and opposed culture in which the dominant principles of universalism and academic freedom created neither an echo for the concept of autarchy nor a justification for submitting research priorities to an agenda of economic/technological development. The technology bureaucracy's corporate culture leaned towards the 'antidependency' approach (Adler 1987) which is described above, while the science bureaucracy's evolved around the question of standards of excellence and is in general more geared to an immediate concern with self-survival.

However, despite the factors that inhibited technology policies in Latin America and the rather sporadic and limited political base from which it customarily operated in these countries, there are some brighter spots. The limitations of the political base serve to explain the fact that there was not much success with building a national innovative system; the few successes suggest that better results may have been possible. There is some room for public inducement of putatively desirable technology developments which would not otherwise have occurred.

Why, then, have Latin American states been relatively successful in some specific sectoral technological policies? Nun's discussion of the rationale underneath the success of INTA in the Argentine agricultural sector is illustrative and may perhaps be generalized. According to him, that agency did not seek to redefine but rather to attach itself to the economic model that traditionally ruled agriculture in Argentina. It also promoted a network of regional centres, experimental stations and extension agencies. Finally, the demand for technology was induced by the creation and continuity of a policy of credits and tax allowances which meant subsidies for capital investment while the supply of technology was mainly promoted internally by INTA. Success in promoting technological change in agriculture may be facilitated, as suggested by Nun, by the peculiarity of innovations required by rural enterprises. Most of these innovations are not likely to be patented or monopolized, so that more of the research efforts involved tend to be external to firms. In contrast, informatics policies of Argentina and Brazil were much less successful (there are some who even think they were a complete failure) as sectoral technology policies. Not only were they unable to stimulate intrafirm technological effort, but also the demand for technology that was stimulated by the very political decision of creating a domestic IT industry was conditioned to the domestic market. However, this market was not prepared to supply the necessary technology at the qualitative level required and/or at a reasonable price. In addition, the informatics policy in both countries, but perhaps more deeply in Argentina than in Brazil, clearly contradicted the interests in the rest of the industrial sector and in the economy as a whole. Industrial customers of the informatics sector continued to want access to foreign suppliers and so found the informatics policy at variance with their interests.

The contrasts seem to contain some lessons. The success of technology policies for innovation in agriculture in Latin America, when compared with other state initiatives in the region, seem likely to have been due to sustained state action and also to a 'style of intervention' which strongly favoured the private process of accumulation and was consistent with the perceived political and economic interests of the private agents involved. They seem to have been based on the identification that what was good for these sectors was also good for the whole economy. It became feasible to convince the producers that what was good for the whole economy was also good for their private gains. Space for efficient public technology policy is created when state guidance converges, adds, supplements general trends, or stimulates the emergence of new ones in production. Or, in more poetic words, the probability of state efficiency in technology policy increases when the state acts less as a custodian (regulator) or demiurge (producer), than when it plays the role of 'midwife' in assisting the emergence of new entrepreneurial groups or inducing existing groups to venture into new sectors (Evans 1993). This seems to be an obvious conclusion, but in relation to technology policy in Latin America it is not: many such policies have been seen and used for a long time as a component of a strategy of resistance, not as a tool to improve the private process of accumulation and, through it, promote technology development and economic growth. In this respect, the biggest obstacles faced by firms and governments in Latin America on the way to competitiveness are less financial or technical but much more institutional and ideological (Perez 1993: 1).

2.3 State capacity of technology policy-making: policy content and responsiveness of technology agencies to change

This topic is well explored in the chapters devoted to technology policy for electronics and informatics. Erber shows that deterioration of the Brazilian IT policy was well under way in the late 1980s, not only because of strong external pressure, but also because of lack of policy co-ordination within a government trying to cope with diametrically opposed views about the policy and living off thin political support. The main reason for the policy's demise was its inability to retain social support – which Erber shows to have been in great part a result of characteristics of the policy itself. The objective of developing local innovation capability under control of national enterprises made the policy intrinsically conflictive in his view. The erosion of support within the informatics entrepreneurial community was felt when many of them started to negotiate partnerships with foreign companies, increased legally and illegally imported content and restrained investments in R&D. This reduced commitment to the policy's objectives and rules may be explained either as a sign of entrepreneurial capability for interpreting market signals and adapting strategies or as an indication that many of them were never really committed to the policy's objectives but supported them as an ideological cover for their rent-seeking behaviour.

The Brazilian policy's technology strategy was built on four critical assumptions which later proved to be false. First, it was assumed that, having mastered the innovation capability for a family of products based on a combination of imported technologies and endogenous investments, local firms would be able to produce without further technology imports. Second, it was assumed that national enterprises would invest the necessary amounts to absorb the imported technologies and develop their own technologies. Third, the speed and intensity of technological change in the international electronics industry was thought to be manageable. It was not. The threshold of resources necessary for catching up was consequently underestimated and the emphasis on product design skills proved less strategic than the emphasis on production and marketing skills. Fourth, consumers were assumed to be able to bear the costs of the policy in terms of restricted product range, higher prices and lower performance of goods supplied. Their dissatisfaction, nurtured by the publicizing of the industry's inefficiencies, exhausted their tolerance and caused a shift, in the late 1980s, to hostility and civil resistance.

The slow reaction of the policy to the transformations of the international electronics industry and the problems facing the policy domestically is pointed out by Erber as one of the important causes of the policy's downfall. The policy's insulation from the academic community (mainly due to the 'national security' origin of SEI) contributed to reducing the agency's capability properly to assess the technical changes at the international level. It was also 'insulated' from society as the agency was in contact only with those groups and organizations which were, in principle, strong supporters of the policy. The policy's inwardness was not a strategy, but was policy-driven: success of internal sales reduced incentives to export; emphasis on local design reduced their tradability; the scant attention to production technology had negative effects for international competitiveness.

The designers of the Argentine electronics policy had a highly relevant example to emulate in this Brazilian experience. The initiation of the Argentine policy coincided with the approval by Congress of the Brazilian policy. This approval received massive domestic political support rivalled only in Brazilian history by that in favour of state monopoly of petroleum extraction and refining, some thirty years before. However, almost one year later, the threat of economic sanctions by the US government, signalled to the Argentine industrialists, certainly much more visibly than the internal weaknesses of the Brazilian policy, the poor prospects of such policy.

A consideration of the effects on Argentine society of the trade conflict between Brazil and the US over the informatics policy, and the influence such a conflict might have had on the Argentine policy, are missing in the otherwise rich analysis of Chapter 5. Nochteff shows us, however, the strong internal inconsistencies in the two central components of the Argentine electronics and informatics policy: the tariff and the industrial promotion schemes. A combination of the late application of lower tariffs on imported components and materials and the failure of granting tax exemption to companies who succeeded in winning bids for industrial production had the result that (in point of fact and contrary to expectation) the electronic capital goods industry, including computers, was one of the least protected industries in Argentina. Besides, the policy was not able to change the traditional behaviour of the public administration, particularly in relation to procurement, nor to change, as already pointed out, the behaviour of the Argentine major economic groups who did not respond fully to the policy, nor of those who did articulate their production and technology strategies as envisaged by the policy.

3 PROSPECTS FOR STATE ACTION IN SUPPORT OF SCIENCE AND TECHNOLOGY IN LATIN AMERICA

The purpose of this concluding section is to address the question: what role does science and technology policy play in future economic and social development in Latin America, and what kind of technology policy is likely to be needed? The question is, in the final analysis, about state action, possibly state intervention, that may be required in view of the role which technological factors are expected to play in the future development of the region. Any attempt to answer it must, to start with, take account of two main considerations.

The first is the obvious fact that there is widespread disenchantment with the results of state intervention in the Latin American countries. This disenchantment is, in part, a by-product of the ideological offensive which has been launched against interventionist policies in the interests of free markets; it also has solid and objective foundations in Latin American experience.

The second is the fact that, however incomplete the recent shift to freer markets may have been in Latin America, it has resulted in the opening of Latin American economies to international trade and the dismantling of the protectionist apparatus of the import-substituting economy. Import substitution in its historic form is dead. The discussion must therefore focus on the role of technology policy in the context of liberalized open economies.

First then, there is the general question of the role of the state. The contributors to this volume share a rather general pessimism about the accomplishments of the state in the large Latin American countries, which probably finds an echo in more widely spread public opinion. This pessimism is not only about the historical legacy in terms of the performance of domestic economies, industrialists' behaviour patterns induced by traditional forms of state intervention, and the characteristics of the state and political processes in the region. It also informs perceptions about the capacity of the state to break the current deadlock of recession and to engineer an improved participation in the international economy. And this pessimism persists notwithstanding a recognition that there have been some improvements in the performance of state institutions: for example, by way of increased accountability of government officials and more regular functioning of the mechanisms of formal democracy in Latin America which could signal a more positive context for policy implementation.

The historic role of the state in the large Latin American countries, and for that matter in the small ones too, is well recognized. There is agreement as to what the Latin American states actually did, even though there may be great differences of opinion about whether things could have been done better. State intervention in the economy in Latin America provided the fundamental basis for industrialization. It resulted in not only a complex set of regulations, but also the direct involvement of the state in production and a major growth of public administration. The characteristics of this state-induced industrialization were not generally seen as major obstacles to economic development until domestic financial crisis and foreign debt made a reconsideration unavoidable. In that process of reconsideration, internal political criticism of state intervention played a part, but it was greatly supported by a strongly reinforced international economic orthodoxy from the international financial agencies, and more to the point, by the hard realities of lending conditionalities.

The outlines of those historical developments are well known and need not be repeated here. For present purposes the important consideration, which must condition future policy, is simply that the collapse of the political consensus underlying the policies of the past, was accompanied by a major dismantling of the apparatus of state intervention which those policies had brought into being. And this, in turn, has changed the nature of state institutions as well as the relationships of the state to society. Those changes are now built into the political context; their implications are discussed further below.

The second general theme which sets the new context for policy discussion about technology, as about nearly everything else, is the fact of liberalization: the opening of the Latin American economies to international trade and to the operation of international prices. Contributors to the volume show considerable scepticism about the inherent merits of the liberalization programmes as they have been conceived up until now. The general view seems to be that the increased competitiveness which liberalization requires would in principle be a positive development, but that its implications need to be more explicitly considered. All would agree that increased competitiveness based on rising productivity could be highly beneficial. There may also be some agreement that reduction of protectionism will provide a necessary impetus for increasing productivity, and for a diversification of the product range. But at the same time there is considerable doubt about the capacity of unaided market forces to deliver these improvements. If there is a consensus on this point, it seems to be that increased competition in international markets is perhaps a necessary condition, but it is unlikely to be sufficient. The contributors in the main see a need for selective but active intervention by the state. This may reflect the influence of the Southeast Asian model – at least in one interpretation.

This scepticism echoes a growing concern about some of the more simple-minded approaches to liberalization that are current, especially those in which the attainment of competitiveness is put forward almost as an objective desirable in itself. Paul Krugman has recently attacked the notion in a particularly vigorous way: 'the idea that a country's economic fortunes are largely determined by its success on world markets is a hypothesis, not a necessary truth; and as a practical empirical matter, that hypothesis is flatly wrong' (Krugman 1994: 30).

And beyond this, it is argued, the benefits of competitiveness virtually depend on how competitiveness is actually attained. Competitiveness based on increasing productivity is central to rising living standards; but it can also be attained by other means – varieties of simple cost-cutting for example – which in so far as they entail reductions of the real wage (either directly through reducing money wages in relation to prices or perhaps by successive currency devaluations) may not be at all beneficial to living standards. In this sense it is easy to agree with the view that national living standards are still overwhelmingly determined by domestic factors rather than simply by competitiveness on international markets. Competitiveness is of course welcome, especially in so far as it provides an injection of effective demand, but if anything it is, a necessary not a sufficient condition for rising living standards.

This view finds a ready echo in the arguments of the contributing authors and underlies their caution about the benefits of liberalization and the dismantling of state policy systems. In Latin America the rhetoric of competitiveness has of course been widely used as a justification for a whole range of 'tough' policies, including painful cuts in public expenditure and privatizations to reduce public deficits. Attainment of international competitiveness is intended to make the medicine easier to take: it sugars the pill. But, in all this there is very little consideration of the terms on which that competitiveness will be achieved and hence of its real implications for the future growth of society's welfare. The risk is that the older rhetoric of the import-substituting strategy has simply been replaced by another newer variety, obviously more in tune with immediate political projects, but without much regard for fundamental economic concerns about rising productivity, increases in the real wage, and improvements in the standard of living. It is by way of a more systematic concern with the terms on which liberalization is to take place and on which competitiveness is to be attained, that it becomes possible to respond to Frischtak's (1993) concern for Brazil: 'to articulate an alternative development project that addresses major concerns of the general political culture: to make integration in the world economy, and a commitment to competitiveness, compatible with the aspirations of reduced vulnerability'.

A straightforward way of putting all this is as follows. International competitiveness may be based on static comparative advantage and may orient industrial production towards sectors where competitiveness is maintained mostly by a slow-growing structure of real wages. Technology policy will not be a relevant issue in this situation and demand for scientists and researchers will be rather limited. The collaborators to this volume agree that this manner of interpreting competitiveness is presently perceived as the choice made in the major Latin American countries. Evidence on recent industrial development from a number of countries seems to support them. They also argue, in tune with the points made above, that this choice has meant an abandonment of development concerns, which have been overshadowed by a preoccupation with the short term. It is a policy package which may easily mortgage the future. It has the implication and the result that the political alliances around technology policies of earlier years have been dissolved.

Any discussion of the future of technology policy in Latin America has to take the considerations set out above as essential initial conditions. The remarks that follow sketch elements for future policies. They start with an extension of the points already made about the state as an agent in the economy. The discussion of the possible roles of Latin American states in future years must take the radical changes in state policy systems and in the bureaucracy which recent transitions have wrought as a point of departure. It is not possible to discuss the 'role of the state' without facing the historical fact that the state has been changed. The powers available to the state today have been considerably diminished by experience. As far as technology policy is concerned, the contributors to the book share the view that the institutions built up over the 1960s and 1970s, which maintained some kind of vitality into the 1980s, are largely defunct. This at least must be seen as a central constraint.

The transitions of the last decade have meant that the range of actions open to the state are considerably more limited than they were. It is useful to consider three aspects, which are likely to influence the shape of technology policies in the future.

First, the state now has access to proportionately fewer resources than in the period of intervention and import-substitution. Quantitatively, this is reflected in the falling proportions of GDP devoted to state expenditures – at least in some countries. But apart from this, the exigencies of sustained fiscal crisis and inflation in the countries discussed here has meant that their freedom to spend is much more sharply constrained than was historically the case. There is, therefore, a strong bias against policies that are costly in terms of direct public expenditure. At the same time, the state is much less of a source of 'goodies' for the private sector or any other part of the society than it used to be.

Second, under conditions of open economy (which seem likely to continue in the countries discussed here), the range of options open to the state in terms of policy action is restricted. The blanket use of infant industry justifications of the past will no longer work: if the state seeks to encourage some learning processes, it will only be able to do so, if at all, by subsidy and not by tariff protection. And perforce, such policies will be selective and will have to be sharply cognisant of the intersectoral inefficiencies that protection can produce, as for example in the informatics policies discussed in this volume.

Third, related to both of these points, it would seem that the relative roles of the state and the private sector have changed. The state has less leverage because it has less resources; it also has less to offer because it cannot provide the ultimate advantages of protection. In the future, state intervention will be based less on the state as cornucopia, and more on sharing agreed economic objectives with private economic agents.

All this points towards a search for 'cheaper' kinds of policies than in the past, in the sense that neither the resources nor the institutional structures of earlier technology policy will be attainable – or politically and socially acceptable. The day of the technology registers and of large-scale state expenditures on R&D, supposedly for industry, are probably gone. Instead, if there are to be interventions, they will have to take the form of co-operation between the private sector and the state of a kind which more fully engages the private interest than in the past – and which are restricted in time. The perpetually infantile industry has gone for ever.

But this begs the question: are interventions needed? It is hard to conclude that they are not. There are three parts to the argument. The first has already been made. It is simply that the medium and long-run consequences of competitiveness essentially depend on the terms under which it is attained. The argument made here is that competitiveness is more consistent with a long-run improvement in social welfare if it is based on a real improvement in value added per worker. This may take the form of consistent process improvements or of shifts towards more sophisticated products. At all events, these advances open the way for rising real wages and fuller participation in the gains from trade – an old concern in Latin America and no less relevant for that. The preoccupation with technological change as the central basis for long-run competitiveness is all the stronger in a world economy which appears to be experiencing generic technological changes and where the once traditional sectors are increasingly drawn into the cycle of innovation. It will be hard to maintain competitiveness without technological change even in traditional sectors. The social implications of trying to do so would be unpleasant and most likely unsustainable politically.

Second, this emphasis on a technologically based approach to competitiveness opens classical concerns about market failures. These arguments are well established in the literature. The learning process, the accumulation of technological capability and the like, are subject to all manner of externalities, and are archetypally matters in which markets are unlikely to operate satisfactorily. Private investments in such areas tend to be suboptimal. The arguments are largely ignored in the present climate of ideological commitment to market economics, but they remain valid. In so far as they are accepted, they imply that the state has a role in stimulating investment in the accumulation of technological capability. They do not necessarily imply that the state should itself make those investments, and a good deal of the history of recent years suggests that it probably should not.

Third, it is nevertheless argued with considerable force and some justification, that the process of technological learning should be left to the market, however fallible it may be. The assertion is that bad markets are better than bad states. This has considerable appeal, though it is a great deal more difficult to establish than its proponents appear to believe. Usually the argument is based on counterfactual 'empiricism'. For example, it might be accepted that the Japanese state of the 1960s and 1970s was quite strongly interventionist in relation to technological factors, but it is asserted that things would have gone even better if it had not intervened at all. Sometimes the difficulty of establishing the argument is simply ignored – see for example the World Bank (1993) study of the 'East Asian Miracle' which contrives to dismiss the role of the state in NIC's economic development, and by and large simply avoids serious discussion of the technological factor through which state action was pre-eminently expressed.

It is highly likely therefore that the economic and social limitations of a pattern of development based on short-run comparative advantages will sooner or later move Latin American governments to the point of intervening in favour of technological advance in industry. That view is certainly consistent with the material presented here. And despite some recent attempts to play down the recent historic successes of state intervention in favour of accumulation of technological capabilities, the evidence stands and can only be refuted by counterfactuality. It remains true that whilst we have cases of state organizations making an inefficient mess of technology policy interventions, we also have cases of apparent successes. Furthermore, we have no examples of a successful, technologically based policy of competitiveness in developing countries in which the state was neutral or inactive. On the contrary, the present Latin American evidence, albeit covering a relatively short period, seems to suggest that when there is no serious intervention in favour of technological change, the economy tends to get stuck in traditional short-run patterns of comparative advantage. These may produce exports, but will they produce development?

REFERENCES

Adler, E. (1987) *The Power of Ideology. The Quest for Technological Autonomy in Argentina and Brazil*, Berkeley, University of California Press.

Bardham, P. (1993) 'Symposium on democracy and development', *Journal of Economic Perspectives*, vol. 7, no. 3, pp. 45–9.

Bradford, C. I. (1994) 'Redefining the role of the state: political processes, state capacity and the new agenda in Latin America', in Colin I. Bradford (ed.) *Redefining the State in Latin America*, Paris, OECD.

Evans, P. (1993) 'Three tales of NICs and computers: reflections on the political dynamics of technology policy', paper presented at the First INTECH Conference, Maastricht, June 21–3.

Frischtak, L. (1993) 'The long transition: autarchy, competitiveness and industrial policy in Brazil', paper presented at the First INTECH Conference, Maastricht, June 21–3.

Haggard, S. and Kaufman, R. (1994) 'Democratic institutions, economic policy and performance in Latin America', in Colin I. Bradford (ed.) *Redefining the State in Latin America*, Paris, OECD.

Huber, E., Rueschmeyer, D. and Stephens, J. D. (1993) 'The impact of economic development on democracy', *Journal of Economic Perspectives*, vol. 7, no. 3, pp. 71–85.

Katz, J. M. (1976) *Importación de tecnología, aprendizaje, industrialización dependiente*, Mexico, Fondo de Cultura Economica.

Krugman, P. (1994) 'Competitiveness: a dangerous obsession', *Foreign Affairs*, vol. 73, no. 2, pp. 28–44.
Lijphart, A. and Crepaz, M. M. (1991) 'Corporatism and consensus democracy in eighteen countries: conceptual and empirical linkages', *British Journal of Political Science*, vol. 21, Part 2, pp. 235–46.
Mowery, D. C. (1994) *Science and Technology Policy in Interdependent Economies*, Boston, Dordrecht, London, Kluwer Academic Publishers.
Perez, C. (1993) 'Technology and competitiveness in Latin America: beyond the legacy of import substitution policies', paper presented at the seminar 'Globalization, Liberalization and Innovation Policy', IDRC, Ottawa, 27–9 May 1992.
Przeworski, A. and Limongi, F. (1993) 'Political regimes and economic growth', *Journal of Economic Perspectives*, vol. 7, no. 3, pp. 51–69.
Reis Velloso, J. P. (1994) 'Governance, the transition to modernity, and civil society', in Colin I. Bradford (ed.) *Redefining the State in Latin America*, Paris, OECD.
Smith, P. H. (1992) 'On democracy and democratization', in Howard J. Wiarda (ed.) *Politics and Social Change in Latin America. Still a Distinct Tradition?*, Boulder, Col., Westview Press.
Sørensen, G. (1993) 'Democracy, authoritarianism and state strength', *The European Journal of Development Research*, vol. 5, no. 1, pp. 6–34.
Streeten, P. (1993) 'Markets and states: against minimalism', *World Development*, vol. 21, no. 8, pp. 1281–98.
World Bank (1993) *The East Asian Miracle. Economic Growth and Public Policy*, New York, Oxford University Press.

Index

Printed in the United States
by Baker & Taylor Publisher Services